VIOLENT WEATHER PREDICTIONS 2000–2001

Forewarned is Forearmed

In a world that inevitably has millions of people dealing with the devastating effects of natural disaster, there is no substitute for advance planning based on advance warning. To this end, science has come up with an impressive array of methods and technologies for measuring and predicting the awesome destructive power of floods and storm surges, volcanic eruptions, fierce winds, and tumultuous earthquakes.

Now, add to these methods the impressively accurate and precise science of astrometeorology. By looking beyond the horizon of our own world into the complex interaction of the other stellar and planetary bodies of our solar system, author Jennifer Lawson has given us *Violent Weather Predictions* 2000–2001, a thought-provoking book that includes an amazingly specific schedule of dates and places where the Earth might unleash its most deadly fury at the turn of the century.

About the Author

Australian-born Jennifer Lawson is a long-range weather forecaster. SInce her first public prognostication (Tropical Cyclone Pierre in 1985), her successful predictions of the dates and locations of such storms have been published every year in the Australian media.

In the United States, where she resides part-time in the Houston area, she has appeared in the local media every year since 1988 with accurate predictions of the dates and locations of each season's hurricanes.

To Write to the Author

If you would like to contact the author or would like more information about this book, please write to her in care of Llewellyn Worldwide. We cannot guarantee that every letter will be answered, but all will be forwarded. Please address your correspondence to:

Jennifer Lawson
c/o Llewellyn Publications
P.O. Box 64383, Dept. K414–6
St. Paul, MN 55164-0383, U.S.A

Please enclose a self-addressed, stamped envelope for reply or $1.00 to cover costs. If ordering from outside the U.S.A., please enclose an international postal reply coupon.

VIOLENT WEATHER PREDICTIONS 2000–2001

Countdown to Cataclysm

Jennifer Lawson

1999
Llewellyn Publications
St. Paul, Minnesota 55164–0383

FIRST EDITION
First Printing, 1999

Cover design: Lisa Novak
Editing and layout: Ken Schubert

Library of Congress Cataloging-in-Publication Data
Lawson, Jennifer, 1945–
Violent Weather Predictions 2000–2001 : countdown to cataclysm / Jennifer Lawson. – 1st ed.
p. cm.
Includes bibliographical references and index.
ISBN 1–56718–414–6
1. Severe storms–Forecasting 2. Long-range weather forecasting. 3. Natural disasters. I. Title.
QC941.L34 1999 99–40795
551.64'5--dc21 CIP

Llewellyn Worldwide does not participate in, endorse, or have any authority or responsibility concerning private business transactions between our authors and the public. All mail addressed to the author is forwarded but the publisher cannot, unless specifically instructed by the author, give out an address or phone number.

Llewellyn Publications
A Division of Llewellyn Worldwide, Ltd.
P. O. Box 64383, St. Paul, Minnesota 55164–0383, U.S.A.
www.llewellyn.com
Printed in the United States of America

Contents

Chapter One

Our Violent Universe

Chaos, as defined by the dictionary, is the confused state before the creation of the universe. By definition, the state of chaos that exists before creation changes to a state of order at the point of creation—which simply means that out of the primordial chaos come orderly patterns. Chaos Theory demonstrates how these emergent patterns are evident but not predictable.

Current scientific models propose that billions of years ago our universe was created in a "Big Bang," an explosion that radiated more energy than a trillion suns, bringing the state of order and the formation of our universe, galaxies, stars, and planets.

The Sun and Solar System

Within our universe, there are billions of galaxies. Each of these is thousands of light years across and contains millions of stars.

Our own galaxy, the Milky Way, is a spiral galaxy, typical of many galaxies where millions of stars rotate around a center. As the mass of stars sweep around this center, long arms of dust and gas form. Over millions of years these gas clouds become so dense they collapse under the pull of their own gravity, and new stars are born.

As gas and dust collapsed through gravitational action, its turning momentum caused ever faster rotation until a point was reached where a residue was spinning too fast to continue collapsing. The collapsed material became our Sun and the so-called "accretion disk" became the planets.

Our Sun is a solitary medium-sized star. It is considered to be among a minority of stars that formed in isolation, for most of the suns or stars in our universe formed in pairs, triplets, or in even greater numbers. A single orbit of the Sun around our galaxy takes something like 225 million years to complete.

If our Solar System were comprised of two or more suns instead of one, any planets forming from the collective suns' debris might experience very unstable orbits and great extremes of temperature, making the potential for life developing under such torrid conditions almost negligible.

More than ninety-nine per cent of the total mass of the Solar System is contained within our Sun, which provides all the light and heat throughout our Solar System, and provides the Earth with a perfect environment in which life forms can flourish.

Our Dynamic, Delicate World

Most of the time we tend to take life on Earth for granted, but life on Earth is really held in a very delicate balance. At any given time a major variation in the Sun's temperature, magnetic field, or gravitational field could threaten life on Earth. If a nearby star went supernova ten light years away, it would be about as bright to us as our own Sun. If our own Sun went supernova, we would be part of that explosion.

The Earth's inner core is composed of iron and surrounded by a liquid molten outer core where temperatures reach a searing 9,000 degrees Fahrenheit. Around the molten core is the Earth's lower mantle where temperatures are a little cooler—some 2,500 degrees Fahrenheit—and separated from the upper mantle by a denser rocky layer. The uppermost mantle and the thin fragile floating outer layer of the Earth's crust, known as the lithosphere, is where we all live.

Around 225 million years ago the Earth was one landmass known as the supercontinent P*angaea*. Around 180 million years ago Pangaea began to break up, and like a seam being slowly torn apart continents slowly but surely parted from each other. Over millions of years, the separated continental plates continued to slowly drift apart.

Around 65 million years ago Antarctica was still joined to Australia, and North America was in the process of separating from Eurasia leaving Greenland in between them.

Tectonic Plates, Continental Drift, and the Pacific "Ring of Fire"

The Earth's lithosphere or "crust" is broken up into gigantic tectonic plates. There are six major rigid plates and several smaller plates. Where these plate boundaries or fault zones intersect, volcanoes are formed and earthquake activity occurs.

There are different types of plate boundaries. These include convergent plate boundaries, where one plate dives under another; divergent boundaries, where new crust is generated when one plate pulls away from another; and transform boundaries, where plates slide past each other in a slip-strike action such as California's San Andreas fault. There are also plate boundary zones where plate action is less clearly defined and more diffuse.

The most active fault zone in the world circles the Pacific Ocean, for it is here that several large tectonic plates meet and jostle for supremacy. Two-thirds of the worlds' volcanoes are located along the margins of the Pacific Ocean giving the Pacific Rim the name "Ring of Fire." Not only is volcanic activity more prevalent here than anywhere else in the world, but the highest concentration of earthquakes also occur around the Ring of Fire.

Of the countries bordering the Pacific Rim, Japan and Mexico experience a predominantly higher amount of earthquake and volcanic activity. These two countries sit atop three large intersecting tectonic plates which are in a constant tug-of-war created by the enormous stresses within the Earth's fragile crust. Japan alone has eighty active volcanoes, sixteen of which are considered to be particularly dangerous.

The North American plate rubs shoulders with the Pacific plate along the West Coast of the USA, and is responsible for California's mighty San Andreas and Great White Wolf faults. The entire North American West Coast is considered a high quake risk area and major "hot spot." Another tectonic plate boundary spans almost 6,000 curving miles from Italy, Greece, and Armenia to Turkey, Iran, and India, finishing beneath Indonesia. Of all these countries, Indonesia is the most seismically active, experiencing constant volcanic and earthquake activity.

Although large quakes strike more often along these tectonic plate boundaries than any other regions of the world, massive earthquakes can and do strike anywhere on Earth—even in seismically inactive regions of the world.

The most destructive, powerful, and prolonged series of earthquakes—unparalleled in the recorded history of the United States, and quite possibly the world—did not occur along California's mighty San Andreas fault zone as one would suspect, but further to the east in the Midwest's quiet, peaceful Mississippi valley.

It was here on December 16, 1811, that a massive tremor (estimated at 8.6 on the Richter scale) rocked the sleepy Mississippi Valley. The New Madrid tremor was felt over one million square miles—almost two thirds of the United States—as shock waves traveled as far north as Montana, as far south as the Gulf Coast, and east to the Atlantic Seaboard. It triggered landslides in the mountains of North Carolina; set bells ringing as far east as Norfolk, Virginia; buckled sidewalks in Baltimore, Maryland; and stopped clocks as far north as Boston, Massachusetts.

This powerful quake directly affected 250,000 square miles. (By comparison, the great San Francisco quake of April 18, 1906, directly affected 12,000 square miles—twenty times less.) Land levels changed by as much as twenty feet, prairies subsided, swamps were drained, farmland was turned into swampland, new land was formed, and two new lakes appeared in Arkansas and Tennessee.The Mississippi River temporarily flowed backwards, generating a twenty-foot wave which literally engulfed hundreds of acres of land, permanently changing the course of the river in many places. Debris from wrecked boats and jetties littered riverbanks and covered the river for miles. Eyewitness reports at the time describe the Earth rippling like waves on the ocean; and large volumes of water, sand, and coal being ejected some forty feet into the air.

For two months, powerful, destructive aftershocks of more than 8.0 magnitude continued to rock the central Mississippi Valley, Finally on February 7, 1812, the last and most powerful tremor of all struck. This monstrous quake completely obliterated the town of New Madrid and neighboring towns as the ground rose in the air and then subsided. For a five-month period following these powerful quakes, sizeable aftershocks continued to plague the Mississippi Valley. The Mississippi Valley recorded more powerful tremors in this short time frame than were recorded in California over the period from 1932–1972. Because the tremors affected such a large area and were so prolonged, the death toll was never fully known.

No place on Earth is really secure from earthquakes!

Earth's Seasons

The Earth spins on its axis and follows an elliptical orbit around the Sun once every year. Thus its distance from the Sun varies throughout the year. The Earth's axis is tilted, so that at different times during its yearly sojourn one of its poles leans toward the Sun. This motion gives the Earth its seasons.

When the North Pole is tilted toward the Sun, the northern hemisphere experiences its summer, and conversely, when the South Pole is tilted toward the Sun, the southern hemisphere experiences its summer. During summer in the northern hemisphere on July 5, the Earth's elliptical orbit is at its furthest point (aphelion) from the Sun. The South Pole's closest point to the Sun (perihelion) occurs on January 4, during the southern hemisphere summer months.

This could be the main reason why southern hemisphere summers tend to be hotter and one reason why skin cancer, induced by solar exposure, is more predominant in the southern hemisphere than in northern climes.

Our Inconstant Sun

The universe in which we live is a violent, unpredictable one, and the Sun is no exception to the rule. Our Sun is a variable, inconstant star which constantly fluctuates and experiences periodic upheavals. The Sun has a magnetic north and south pole much like the Earth does, but the strong magnetic fields on the Sun are not uniform, and are linked to sunspot activity.

When sunspots reverse polarity roughly every twenty-two years—a period of two eleven-year maxima cycles—the solar magnetic field reverses accordingly. Although the major sunspot cycle peaks roughly every eleven and twenty-two years, there are also other sunspot cycles which occur over a longer or shorter period. Whenever two of these cycles peak simultaneously, the effects are magnified and fierce magnetic storms erupt on the solar surface.

Sunspots and Solar Flares

In 1946, John Henry Nelson, a shortwave radio expert and amateur astronomer, was assigned by RCA to study sunspots to see if they were responsible for shortwave radio interference. Equipped with a six-inch refracting telescope on a makeshift rooftop observatory on the RCA building in New York City, Nelson began studying the Sun.

Nelson observed through his telescope that the sunspots had a magnetic polarity, and that a major sunspot grouping in the Sun's northern hemisphere was always consistent with an opposing identical group in the southern hemisphere. He also noted that sunspots grouped in either hemisphere had a magnetic polarity of a north or south-seeking magnetic pole. When one sunspot cycle died out and a new sunspot cycle began, the magnetic polarities would automatically reverse to the major sunspots grouping in the opposing or southern hemisphere with another "copycat" grouping then forming in the northern hemisphere.

Nelson also observed that, when viewed from a favorable angle, he could see a "nick" or half-oval depression on the edge of the Sun. This indicated that the sunspots were actually holes in the Sun's outer gaseous layer, which looked to him like the swirling formation of a tornado, probably a vortex caused by swirling gas.

From his observations, Nelson deduced that the planets were responsible for sunspot activity, as sunspot activity correlated with planets either in forming magnetic angles to, or eclipsing each other, in their orbits around the Sun. From his findings Nelson began to predict weather on a long-range basis for RCA and established a track record of ninety per cent accuracy over many years. Nelson used heliocentric (Sun-centered) positions of the planets as opposed to geocentric positions (Earth-centered) in weather prediction.

Nelson's hypothesis that the planets directly affected the Sun or indirectly affected Earth's weather patterns came under fire from meteorologists and astronomers, who believed that planets were too far away and their effect too insignificant to matter.

Nelson agreed with his critics that the planets had little effect on the main body of the Sun. He did, however, believe that the planets had a major bearing in affecting the solar atmosphere—where the very unstable, electrified area of the Sun's surface could easily be influenced by the minuscule gravitational tug from the planets. Such influence could create an avalanche effect, triggering major turbulence in the solar atmosphere, subsequent solar storms, and sunspots.

Major sunspot activity and ensuing solar flares send electromagnetically charged particles and solar winds into outer space. Upon reaching the Earth, the solar wind interacts with the upper atmosphere and tears up the ionosphere, triggering electromagnetic or electrical storms on the planet.

The surface of the Earth is protected from direct solar wind and solar cosmic rays by the atmosphere as well as the Earth's magnetic fields, where some of the particles may be trapped temporarily in the Earth's magnetic field or deflected toward the poles.

In *Our Threatened Planet*, Joseph Goodavage talked about unusually powerful solar flares that occurred in early

August 1972, at the lowest ebb of the eleven-year sunspot cycle. The Sun suddenly began to erupt in a series of violent flares and magnetic storms that astronomers (at that time) classified as the most severe ever recorded.

There was even some speculation at that time about a possibility that the Sun was changing polarity. The magnetic storms were of unprecedented intensity and severely affected the Earth's upper atmosphere, knocked out radio communications, and produced a magnificent display of the aurora borealis which was clearly visible over the United States as far south as Richmond, Virginia.

Scientists around the world were baffled by the unprecedented intensity of these extraordinary solar flares. Even X-ray sensors ran right off the scale. Something strange was definitely happening to the Sun.

Supernovae: Spectacular Cataclysms of Nature

Three years later, on July 2, 1975, another massive flare, equivalent to the first, erupted on the solar surface. Not too long after this large solar flare had erupted on the Sun, a star located about 1,000 light years away blew itself apart and flared into spectacular brilliance. Supernovae are exploding stars that flare up to hundreds of millions of times their normal brilliance and, literally, blow themselves apart obliterating anything and everything within their local environment.

An exploding star is not such an unusual event, and in a typical galaxy supernovae occur on average once every hundred years or so. At times, the force of these great supernovas have been so powerful and so brilliant that they have been clearly visible to the naked eye during the daytime.

Following the 1975 solar flare when a star exploded some 1,000 light years away, the scientists thought that the two events may somehow be related.

"By the mid 1970s," said Goodavage, "spectroheliographic analysis of the Sun revealed that almost half of the Sun's store of hydrogen had already been depleted." The Sun was actually burning up at a far faster rate than was previously supposed.

Until the advent of the space age, the Sun was considered to be constant and stable. Now we know otherwise. The Sun is a variable star which flickers, fluctuates, and changes constantly.

In December 1995, an international collaboration between ESA and NASA launched the space-based satellite SOHO to study the solar surface. From its vantage point, the SOHO observatory has helped revolutionize our understanding of the Sun and its effects on the Solar System.

Solar Variability and Global Warming

Global warming, increased levels of carbon dioxide in the atmosphere, melting ice sheets and glaciers, and an increase of severe El Niños all point to a rapid change in Earth's climate.

Scientific opinion is not all agreed that global warming is solely the result of man-made heat-trapping gases.

According to astrophysicist Sallie Baliunas of the Harvard-Smithsonian Center for Astrophysics, the Sun's variability is the greatest factor in global climate change. She claims that increased greenhouse gases in the atmosphere would be slow and insignificant compared to natural climate changes spurred by the sun.

The Earth and the Solar System are interactive, interrelated, and interdependent, and it only makes sense that any fluctuations on our local star, our life-giving force, would play a major role in Earth's changing climate.

An article featured in the January 1999 issue of *Discovery* magazine stated that the Earth had recently slowed down due to the El Niño effect. "As westward trade winds slackened off and eastward jet streams sped up, El Niño made the atmosphere move faster in the direction of Earth's rotation. The total momentum of Earth and its atmosphere is fixed, so in compensation for the strong wind, the planet's rotation had to slow down. When the atmosphere speeds up eastwards the Earth's rotation slows down."

Although the Earth's rotation has been slowing down by about 1.4 milliseconds every century, scientists have recently discovered that it can happen more rapidly due to forces of atmospheric motion, as well as movements within the Earth's molten core, melting polar ice, and ocean currents. In other words, climate plays a major role even in affecting, however minutely, the rotation of the Earth.

What Is Happening to Our Universe?

On December 14, 1997, a massive explosion lasting two to three seconds was observed at the edge of our known universe and was estimated to be second only in magnitude to the original Big Bang that formed our universe around 15,000 million years ago. The gamma-ray bursts of such high-energy radiation had scientists completely baffled. Many more explosions of equal magnitude have been continuing in sporadic bursts ever since 1997.

Wondering what these unprecedented sporadic explosions at the edge of the universe are all about has led astronomers to speculate that we could be witnessing the birth of a new universe, the collision of neutron stars, or maybe even a developing black hole.

Recent photographs of Jupiter shot by spacecraft indicate that the massive storms on the Jovian surface have also begun reversing their flows and are now colliding into one another. Could this phenomena be caused through recent comet fragments from Shoemaker-Levy 9 violently crashing deep into the gaseous giant? Or something else?

Signals Detected from the Milky Way

Within our violent universe and at the very heart of our own galaxy, the Milky Way, new and "almost deliberate" signals have been detected coming from that region since 1991 and are still continuing today.

Also, intense amounts of radiation pulsating from a star twenty light years away, have been bombarding the Earth's atmosphere. In August 1998, a particularly strong burst of radiation pulsing from this star was powerful enough to knock out two satellites. This star, known as SGR1900+14, is a member of the newly discovered class of stars called magnestars, which are purported to be fast-spinning neutron stars with enormous magnetic fields.

Meanwhile, on the solar surface, fierce magnetic storms and solar flares have been raging and bombarding the

Earth with such unprecedented intensity that power grids, short-circuit satellite electronics, and land-based communications have been severely affected. These fierce solar storms could also be partly responsible for a reduction of the Earth's ozone layer.

British astrophysicist Dr. John R. Gribbin, author of many scientific books and co-author of *The Jupiter Effect*, wrote about a correlation between large solar flares and subsequent ozone depletion in *The Hole in the Sky*. In this book Dr. Gribbin refers to research carried out in 1975 by Donald Heath and Arlin Kreuger of NASA based on old data obtained from the weather satellite Nimbus 4. Apparently, the data had never been previously analyzed to correctly show ozone fluctuations following a large solar flare that occurred in August 1972. Using the latest technology, NASA scientists found that the data from Nimbus 4 indicated that a decrease of 16 per cent in the north polar ozone had taken place immediately following this powerful solar flare.

Large solar flares may not only be responsible for a depletion of ozone, but possibly the supercooling in the upper atmosphere as well.

Dangerous Earth-Crossing Asteroids

Perhaps the greatest threat to life on our planet is a chance encounter with a large asteroid, like the one which wiped out the dinosaurs and virtually every existing species on Earth around 65 million years ago.

The largest asteroid belt—some 300 million kilometers wide—suspended between Mars and Jupiter is estimated to be either the remnants of a planet that failed to develop, or a planet that may have disintegrated due to Jupiter's immense gravitational field or a collision with a foreign body. Although the asteroid belt between Jupiter and Mars is the largest in our solar system, it is not the only asteroid group that presents a threat to life on Earth. The Aten and Apollo asteroid groups are of particular concern, as they have orbits that frequently cross that of the Earth. The objects in the Amor asteroid group are not as yet considered to be a threat, for they currently do not have Earth-crossing orbits. However, in future years when their paths evolve and extend out further into space, the Amor asteroid group could also develop Earth-crossing orbits and become a major concern.

There is increasing evidence that some of the objects that we think are asteroids are in fact dormant comets, occupying a Halley-type orbit. When a comet becomes gaseous or burnt out, it loses both its tail and coma and is often mistaken for a dark asteroid.

Although there are currently around 250 known Earth-crossing asteroids, ranging from a few meters to the largest, 1620 Ivar, which has a diameter of nine kilometers, the true figure is probably closer to 2,000.

The asteroid that is thought to have wiped out the dinosaurs is estimated to have been around ten kilometers in size. Judging by the size of ancient craters on the Earth and the Moon, astronomers are of the opinion that asteroids of this dimension have targeted the Earth once every few tens of millions of years. With the advancement of the space age, extensive evidence of asteroid impacts is clearly visible on all of the stable surface bodies throughout the solar system. On Earth alone, more than

150 structures, some masked by heavy erosion and others buried deep under ocean sediments, have now been detected from space and been positively identified as impact craters. With every year that passes, more and more craters are coming to light.

More recent is the discovery of what may possibly be two more asteroid belts; one between Mercury and the Sun, and the other between Earth and Mars. This brings the odds of an asteroid collision with the Earth to a much higher ratio than originally supposed.

Our Restless Earth

From geological evidence found by analyzing the direction of magnetism in rocks dating back 76 million years, the polarity of the Earth's field has reversed polarity 171 times.

More recent geological events and climate changes dating back 160,000 years can now be determined through ice cores. These are drilled by hollow rods thousands of feet into the ice layers in Greenland and Antarctica. By analyzing tiny air bubbles containing carbon dioxide trapped in the ice cores, researchers have found a correlation between Earth's climate changes and carbon dioxide levels. When the climate was colder, CO^2 levels were lower; when the climate was warmer, carbon dioxide levels soared.

Climate changes were not always gradual. Some were very fast and fierce, and occurred within decades. With the sudden onset of a global warming phase, ice sheets and glaciers began to melt, giving way to the invasion of icebergs which destabilized the large tectonic plates, generating violent earthquakes and volcanic eruptions.

Greece's Santorini Volcanic Eruption

In 1538 B.C. the Santorini volcano, located near Greece in the Mediterranean Sea, blew itself apart in a violent explosion. It was estimated to be 3,000-5,000 times more powerful than the fierce eruption of Mount St. Helens of 1980. It sent ash raining down 500 miles away on Crete, and as far south as Egypt.

It erupted again in 1940, and although this eruption was very small by comparison, it was effective enough to generate a ten-foot-high tsunami surging into a small village on Crete. Eyewitness Dr. Colin McDonald at Athens University told of the seabed being exposed shortly before the tsunami came rolling into the village.

If a small eruption can send a ten-foot seismic wave crashing into shorelines 500 miles away, what effect did the massive explosion of the Santorini volcano have in 1538 B.C.?

Searching for clues, archeologists recently uncovered pumice from the Santorini volcano on Crete, dating back 3,500 years. Similar archeological digs in Egypt uncovered thick volcanic ash layers corresponding to the same time period.

Not only did all the jigsaw pieces fit perfectly with the explosive Santorini volcanic eruption, but may have also coincided with one of the most important historical events ever to take place, and one which was to change the course of history.

Santorini and the Crossing of the "Reed" Sea

Although the timing of the Exodus was originally thought to be somewhere around 1250 B.C., the recent archeological discoveries indicate that in all probability it occurred 288 years earlier than previously supposed—most likely around the same time as the powerful Santorini volcano eruption, which would account for the extraordinary events and phenomena surrounding that event.

Moses led the Israelites out of Egypt with the sanction of Pharaoh Rameses II, who had become fearful of the great plagues that had recently befallen Egypt, the last of which had taken the life of his own first-born son. The grieving Pharaoh interpreted these as a punishment by this ominous foreign God of Israel for keeping "His" people in bondage. Rameses II consented to freeing the slaves so that they could return to Canaan under the leadership of Moses, who had been reared in Pharaoh's household.

For some time prior to the massive Santorini volcanic explosion, smaller eruptions had rained down ash over Egypt, as evidenced by volcanic ash layers found in recent archeological digs. This would also account for the days of darkness and the resulting plagues throughout the land. The final plague in which every first-born Egyptian child died is unlikely to be a result of fine volcanic ash, which is not selective in its victims. However, the term "first-born" may also have been mistranslated from "newborn." In that case, fine volcanic ash penetrating the lungs of newborn infants could most certainly have resulted in their deaths.

Biblical scholars have recently discovered that the original Hebrew translation of the "reed sea" was incorrectly translated into Greek as the "Red Sea," and that the famed Exodus crossing actually took place at the "reed sea," and not further south at the "Red Sea," as previously thought. Swampland and reed lakes were once a feature of the landscape from north of Suez to Port Said, near the Mediterranean Sea. It was at one of these that the Israelites are now thought to have made their crossing, which seems more in keeping with the natural sequence of events that led up to and followed the Exodus.

Thinking that the plagues would probably disappear once the Israelites left, the Pharaoh's complacency turned to wrath when the volcano continued to rain ash (plagues?) down over Egypt. Believing that Moses had tricked him, the Pharaoh summoned his armies.

The fierce explosive blast from the Santorini volcano generated a powerful tsunami, possibly 100 feet high, which roared around the Mediterranean sweeping coastlines clean—including Egypt. Prior to the massive wall of water targeting the coast, the marshes or reed seas would have drained of all water, affording the Israelites safe passage to the other side. In all probability, the Israelites arrived at the "reed sea" just as the Santorini volcano began to blow itself apart in a series of violent explosions. With the first violent volcanic blast (the strong east wind—Exodus 14: 21), the marshlands suddenly began to drain, leaving dry reed beds that the Israelites used to escape from the descending Egyptian army.

Arriving at the river bank with the Israelites now in sight on the other side of the "reed sea," the Egyptian army with

Pharaoh at the helm saw that the reed beds were dry and drove their chariots at full speed into the marshes. The wheels of their great chariots became so clogged with mud the chariots became immobilized and stuck in the quagmire.

When the Santorini volcano finally blew itself apart, the fiery blast was seen by the Israelites as "a pillar of fire and of cloud" (Exodus 14:24), and the powerful tsunami (estimated to be traveling at 400 mph), already on its way from the earlier blast, swept in from the Mediterranean Sea and drowned the Egyptian army. The Israelites, unaware that a volcano 500 miles away had blown sky high, interpreted the series of events as signs from God.

Either Moses was psychic, or there was indeed divine intervention for the Israelite crossing of the "reed sea." In any case, it couldn't have been better timed.

Plato's Account of Atlantis: 11,000–12,000 BC

Historians of antiquity such as Democritus, Eudoxus, and Plato gathered remnants of information which told of a global catastrophe or series of catastrophes that wiped out scientifically advanced civilizations around 11,000 to 12,000 years ago. The destruction was so great that all traces of civilization sank beneath the ocean waves, leaving nothing standing as a reminder of what was once "the golden age of science and knowledge."

In *Our Threatened Planet* Joseph Goodavage refers to Plato's *Phaedo*, in which Plato told of antediluvian civilizations around the Mediterranean that knew of other countries, aware that the Mediterranean area occupied only a small portion of the planet. Plato also spoke of the most ancient periods of Egypt's history. He considered Egypt in those times to have been far more advanced than in all of her subsequent dynasties, implying that the Egyptian culture was preceded by a superior ancient civilization which disappeared after what might have been the worst disaster of all time.

That superior ancient civilization, according to Plato, was Atlantis. The kings of Atlantis controlled and ruled many colonies over large areas of what Plato referred to as "the great opposite continent." That great opposite continent, claims Goodavage, was America." If there was no Atlantis then how did Plato learn about the American continent?"

If the Santorini volcanic explosion could send a 100-foot-high tsunami around the Mediterranean Sea as it did in 1538 B.C., Atlantis could have also suffered a similar fate around 11,000–13,000 years earlier, or been wiped out through a small asteroid striking that particular region of the world.

Plato was a highly respected Greek scholar, but modern science dismisses his account of Atlantis as myth on the basis of a lack of physical evidence. Noah's account of the great flood is also dismissed on the same grounds, even though the event is well documented in the Bible and other nonbiblical ancient literature such as the Sumerian epic of Gilgamesh.

Flavius Josephus and the Antediluvians

The antediluvians (Atlanteans?) knew all about the biblical deluge well before it happened and in fact predicted it, according to Flavius Josephus (37–100 A.D.), a Jewish historian and general involved in the Jewish revolt against the Romans at Masada. In Alfred John Pearce's *The Text-book of Astrology* (AFA, 1970) there is a detailed account of what Josephus had to say about the great flood and the ancient civilizations extant before the event.

"Josephus states that astrology was practiced by the antediluvians, who had it from Adam, who received his information from God himself. He further states that Seth, having received instruction to its principles from Adam, and foreseeing the flood, engraved the rudiments of the science upon two permanent pillars of stone; and that the remains of those pillars, he 'Josephus' himself had seen. He also relates that the science was taught by Enos and Noah, who preserved it to the days of Abraham."

Burning the World's Ancient Libraries: Destruction of Records of Antiquity

Whether Atlantis ever existed will never be known because of the destruction of ancient records. At one time all records of antiquity relating to science, art, literature, mathematics, philosophy, medicine, as well as records of the great catastrophic events that befell the antediluvians, were safely stored in various libraries around the world. Throughout the ages most of the world's great libraries and ancient manuscripts were destroyed either inadvertantly or deliberately by ignorant rulers who severed forever our link to the ancient past.

The greatest tragedy of the ancient world occurred when the largest library in the West—the Alexandria library in Egypt—comprised of the Serapeum, containing 30,000 ancient volumes and the Bruchion, containing some 40,000 volumes—went up in smoke when the library was razed accidentally during Julius Caesar's occupation of the city in 48 B.C.

By A.D. 391, the Romans had converted to Christianity under Constantine and Theodosius was the new Christian Emperor. The library of Alexandria had been rebuilt and now contained copies of the ancient manuscripts, as well as carefully pieced-together volumes that had been salvaged from the great conflagration some 400 years earlier.

The ancient manuscripts and studies at the Serapeum library were no longer in keeping with Christian thinking, and in an attempt to stamp out all paganism Theodosius ordered the burning of the Serapeum. Most of the precious manuscripts were destroyed. The remaining manuscripts from the Alexandrian library were finally destroyed by Arab conquerors, who used the ancient parchment as fuel for the city bathhouses.

It is tragic to think that our only link with the past can only be found in rocks and ice, myth and legend. We don't know a whole lot about the unpredictable universe we live in, and it would be unwise to dismiss Plato's account of Atlantis or accounts of the biblical flood as pure myth. If the remains of Atlantis lie entombed somewhere at the bottom of the ocean floor, we can get some idea as to the magnitude and the immense scale of this great cataclysm that only left rock, bones, and stones in a watery grave as silent witnesses to this terrible event.

The Cataclysmic Picture

Could we, in our own lifetimes, be faced with the same fate? What if a sizable asteroid impacted with Earth and wiped out our civilization, leaving not one scrap of evidence behind of what once was a highly-advanced civilization? Scientists thousands of years from now would probably not believe that we had once existed either.

But what do we really know? It took Christopher Columbus to prove the Earth was round and not flat. A century later, when Copernicus dared to say that the Earth rotated around the Sun and not vice versa, he was ridiculed by the religious and scientific authorities of the time.

In the 1970s, most of the scientific community branded Immanuel Velikovsky, author of *Ages in Chaos*, *Worlds in Collision*, and *Earth In Upheaval*, as a crackpot—some even tried to have his books banned when he dared to question the prevailing belief in a stable, evolutionary planet and cosmos, postulating instead that we live in a violent, unpredictable universe.

Recently, though, modern astrophysics has blown some mighty big holes in Darwin's and Newton's theories of a tranquil, uniform cosmos. The more we discover about our universe, the more Velikovsky's theories are proving to be correct, and the better we appreciate the fact that we live in a somewhat cataclysmic universe. In particular, discoveries that the Earth has had a long and violent history of collisions with other bodies have revolutionized previously held concepts and caused one of the most important paradigm shifts in scientific thinking.

In *Earth in Upheaval*, Velikovsky vividly describes cataclysmic events that have occurred at varying intervals on Earth. In what could well be a reference to Atlantis, Velikovsky wrote:

"... myriad upon myriad of living creatures came to life on this ball of rock suspended in nothing and returned to dust. Many died a natural death, many were killed in wars between races and species and many were entombed alive during great paroxysms of Nature in which land and sea contested in destruction. Whole tribes of fish that filled the oceans suddenly ceased to exist; of entire species and even genera of land animals, not a single survivor was left.

"The Earth and the water without which we cannot exist suddenly turned into enemies and engulfed the animal kingdom, the human race included, and there was no shelter or refuge. In such cataclysms the land and sea repeatedly changed places, laying dry the kingdom of the ocean and submerging the kingdom of the land."

The thought that one cataclysmic brush stroke of nature's hand across the canvas of life—wiping out several billion people in a twinkling of an eye—could actually happen is almost inconceivable. It sounds like a scene out of some Hollywood disaster movie. Yet catastrophes of this magnitude have happened many times throughout the history of the Earth, and have also been predicted to happen again.

"For the windows of heaven are opened, and the foundation of the Earth tremble. The Earth is utterly broken, the Earth is rent asunder, the Earth is violently shaken. The Earth staggers like a drunken man, it sways like a hut; its transgression lies heavy upon it, and it falls, and will not rise again."
(Isaiah 24: 18-20)

If a cataclysm of this magnitude occurred today with half the world's population living around the world's coastlines, and millions more living in dangerous geological "hotspots," the consequences would be unimaginable.

Today, with sophisticated space technology we can watch the skies for any large asteroids or meteorites that could cause a catastrophe of such magnitude. To date, nothing of any consequence has shown up. An asteroid originally thought to be a possible threat to the Earth in the year 2028 will in fact miss us by about 600,000 miles—not some 30,000 miles as originally supposed.

In the immediate future, however, we do have something to be concerned about—something that will bear close watching. A major planetary alignment on May 5, 2000, will be occurring at a time of higher-than-normal sunspot activity. This could well trigger major earthquakes and eruptions involving currently active volcanoes that have not recently had a major eruption. These include Italy's Mount Etna, Mexico City's Popocatapetyl, and Montserrat in the Caribbean, just to name a few.

Principles of Astrometeorology

Astrometeorology is a science that seeks to forecast weather based on the angular relationships between the Sun, Moon and planets. In this, one of the most important factors is the declination (the angular distance north or south of the equator) of the Sun, Moon, and particularly the planets. Declination points of the celestial bodies are just as important as longitude and latitude when determining long-term weather patterns.

For example, when two or more of the slower-moving outer planets are grouped together at their most southerly point of declination (23° south at the Tropic of Capricorn), the winter months in the northern hemisphere will be severely cold and summers cooler than usual, while the southern hemisphere summers will be extremely hot and dry with milder winters.

Alternatively, if planets are grouped at their most northerly declination at the Tropic of Cancer (23° north), the northern hemisphere will experience hot, humid summers and warm winters, while the southern hemisphere will experience cooler winters and milder summers.

Several planets grouped around the Earth's equator at zero degrees of declination appear to greatly disrupt and disturb the atmosphere, generating powerful storm systems. The influence of the planet Mars on the equator is perhaps the most effective planet of all in stirring up great storms, hurricanes, tropical cyclones, and typhoons. Fierce hurricane Hugo, which I accurately predicted would devastate the South Carolina coastline on September 20–22, 1989, was an example of Mars on the equator at zero degrees of declination.

How severe a season will be is determined by the declinations of the slower-moving "outer planets," conjoined by the faster-moving planets Mars and Mercury. Because Mercury moves so fast and rotates around the Sun roughly four times a year, its influence on terrestrial weather patterns is more pronounced and brings dramatic changes in weather faster than all of the other planets put together.

The planets, Sun, and Moon act like characters on a stage. First, the outer planets (Jupiter, Saturn, Uranus, Neptune, Pluto) set the scene, then Mars coupled with the inner planets Venus and Mercury enter the stage as the major players. Lastly, the Moon makes the grand entrance on stage, as the final "trigger" to them all.

CHAPTER TWO

Tumult in 2000

The year 2000 could be one of the worst years for natural disasters ever recorded in modern times, as floods of biblical proportions, storms, tornadoes, hurricanes of great intensity, severe weather patterns, powerful earthquakes, and volcanic eruptions wrack the planet. Among the countries hardest hit will be India, the United States, Mexico, Japan, Indonesia, the Philippines, Iran, Turkey, Italy, and the British Isles.

This chapter will be devoted to discussing when and where the major events for the year 2000 are likely to occur. The following dates cover specified time periods when the influences of the planets will be the strongest over that particular region of the world, resulting mostly in severe weather patterns or, in seismically active regions, possible seismic activity.

Although none of the following predictions come with a gilt-edged guarantee that they will surely happen, the specified dates are for those time periods when conditions are ripe for a natural disaster, representing the greatest potential for something to happen. In other words, if disasters are going to occur, they are more likely to occur around dates specified for activity than at any other time. Once the predicted event occurs within the given time frame, the influence is then finished.

Earthquakes and, in particular, volcanic eruptions are by far the most challenging area of all in the line of predictive work, and far more difficult to predict than weather forecasting. For one thing, a powerful tremor is dependent on just how great the stress build-up is in a certain location. Another factor is that the epicenter of a quake may be 100 miles away from where its point of greatest impact or destruction is felt.

The dates listed for seismic activity are for those times when the potential for seismic activity is stronger than normal, and doesn't necessarily mean that an earthquake will definitely strike that area. If smaller tremors are experienced prior to a specified date, then a larger quake could be in the cards around the appointed time.

The main intention of this book is to forewarn and prepare the public for events that are likely to take place. It is not possible to predict every single natural disaster or severe weather pattern that comes along, therefore there will be many disasters that will happen that have not been predicted. Additionally, not every predicted event will happen either. However, it is far better to be prepared for something that may not happen than not to be prepared for something that will!

Severe Weather, Earthquake, and Volcanic Activity: Spring 2000

Although the months of May, October, and possibly early August will be the worst for earthquakes and volcanic activity, mid-March through April could also be quite active. There are also other times during 2000 when seismic activity may occur, though not as powerfully.

From March 16–30 the heliocentric (Sun-centered) positions of Jupiter, Saturn, and Mars, close together in the constellation of Taurus, are already moving into position for a major alignment in May, and are at a right angle to Uranus. This planetary configuration will bring earthquakes, volcanic eruptions, and severe weather conditions.

From April 5–22 the geocentric (Earth-centered) positions of Jupiter, Saturn, and Mars will begin to move closer together in geocentric longitude, culminating in a significant aspect around April 18–22, which will generate powerful tremors and volcanic activity.

Many regions around the world can expect freak storms, deluges, flooding, and powerful winds around these dates. There is a good chance of severe tornado outbreaks in the American states of Texas, Nebraska, and Kansas. Other major events are as follows:

New Madrid Fault and South Island of New Zealand: Tremors, February 4–10

The solar eclipse on February 5, 2000, falls at 16 degrees of Aquarius, in the same degree as the planet Uranus. This combination could bring mild tremors for two regions of the world that may be directly influenced by the solar eclipse—namely, the South Island of New Zealand and the Mississippi Valley's volatile New Madrid Fault.

U.S.A., Africa, Turkey, and China: Earthquakes, March 16–April 22

Although May is the predominant month for major quakes, the time period between March 16–April 22 is also another very active period for quakes and volcanic activity in several geographic areas. Powerful tremors could be experienced in the U.S.A. around central Montana to northwest Wyoming, and possibly in Yellowstone National Park, between March 16–April 22.

Alaska's Kodiak Island or city of Anchorage could experience powerful tremors around April 4–20. Similar events are likely in northeast Turkey or northwest Iran. A major tremor could rock northeast Algeria between March 10–24.

Strong tremors could be experienced in several Chinese provinces in northeast China, as well as southern China near the island of Hainan Dao. May 6–14 could also bring very powerful tremors to southeast China or Taiwan.

Central Chile: Strong Tremors, April 10–22

During this period, powerful tremors could rock a region from central Chile to the Argentine border. Panama could be affected as well.

Northeast Turkey, Northwest Iran: Earthquakes, May 6–21

If seismic activity begins to plague this region in April, a powerful tremor may target the area in May at a time when the influences are stronger.

Southern Italy and New Zealand: Volcanic Eruptions, April 15–May 22

Italy's volcanoes Mount Etna and Stromboli, along with Ruapahu volcano in the North Island of New Zealand, could spring to life around this time. In March 1999, during its latest eruptive phase, gigantic Mount Etna (towering 10,000 feet on the island of Sicily) began to show some indication that a major eruption may be imminent. If Mount Etna has not erupted on a major scale before April or May 2000, there is a strong likelihood that it could do so then.

Further north lies the island of Stromboli where the less volatile Stromboli volcano regularly puffs away in small sporadic outbursts. Although the April–May period may activate Stromboli somewhat, this "baby" volcano doesn't present the threat its unpredictable, explosive "big brother" Mount Vesuvius does.

Located further to the northeast near the Bay of Naples, Mount Vesuvius is currently dormant and is not displaying any signs of springing to life any time soon, but watch for any rumblings from Vesuvius—or any other volcanoes for that matter—around the March–May period, as this will be particularly volatile period when several of the world's volcanoes could be activated. Major tremors could also rock southern Italy around May.

Strong tremors may also be felt in the North Island of New Zealand, along with possible volcanic activity from Ruapahu.

The Great Alignment of May 5, 2000: Catastrophic Quakes and Floods

On May 5, 2000, a rare planetary alignment with six planets almost in a straight line will occur in geocentric (Earth-centered) longitude. The Earth will be aligned on one side of the Sun, while the planets Mercury, Venus, Mars, Jupiter, and Saturn will be aligned on the other.

The planet Pluto will also be on the same side as the Earth, though offside and not featured as strongly as the other planets in this powerful alignment. However, the other two planets, Uranus, and Neptune, will be directly involved, making an approximate right angle to the grand planetary alignment.

According to a report written by Peter Michalski in the German newspaper *Bild* (May 6, 1998), the last time that six planets were in a similar alignment was some 6,000 years ago, in 4,000 B.C. From geological evidence dating back to that time period, there are indications that floods, earthquakes, and volcanic activity targeted the Earth. The report also referred to another planetary alignment that occurred in February 1962 in which five planets were similarly aligned in geocentric longitude. The planets Mercury, Mars, Jupiter, and Saturn were aligned on one side of the Sun, while the Earth was aligned on the other. The February 1962 configuration, according to Michalski, triggered a major quake in Iran which killed 20,000, as well as killer floods that swept through Europe, taking some 600 lives.

Another planetary alignment in 1813 (this time with only four planets involved) coincided with a year without summer in Europe. For months on end, the mercury dipped well below the freezing point. Severe crop losses were experienced throughout Europe, resulting in famine and starvation.

English astrophysicist Dr. John Gribbin and Washington-based astronomer Dr. Stephen Plagemann talked about planetary cycles in their book *The Jupiter Effect*, published in 1974. Studying the planetary alignment cycles, Drs. Gribbin and Plagemann observed that when all the planets lined up with Jupiter on the same side of the Sun (around every 179 years or so), great earthquakes ensued.

"The combined pull of the planets stretch the solar surface," said Gribbin, "increasing sunspots and causing powerful solar activity, where resultant streams of charged particles flow out past the planets, bombarding the Earth's atmosphere in the process. Resultant movements of large masses of the Earth's atmosphere then agitates regions of geological instability, causing earthquakes."

Based on their research, Drs. Gribbin and Plagemann believed that when the 179-year cycle peaked on March 10, 1982, (a time when all the planets would once again be aligned on one side of the Sun, with Jupiter close to Saturn) a major earthquake would strike the west coast of the United States.

Popular science magazines and newspapers at the time depicted the planets in a straight line like a row of pearls, but this was not so. In March 1982, the planets were not in a straight line at all, but rather scattered across eighty-five degrees of the heavens in heliocentric (Sun-centered) positions. This is almost one fourth of the zodiacal circle, nowhere near as tight as the closely-knit group will be on May 5, 2000.

Although there was no major quake on the West Coast to speak of, 1982 was one of the worst years on record for weather catastrophes, coinciding with one of the most intense El Niños recorded this century.

Planetary alignments however, do not always coincide with major events such as earthquakes or weather disasters. The effects of a planetary alignment is largely dependent on how tight the grouping is, how many planets are involved, and what angles the planets are forming to each other at that time.

In reference to the May 5, 2000, planetary line-up, Dr. Jacqueline Mitton, a spokesperson for the Royal Astronomical Observatory in London and an expert on the Catastrophe Theory, was reported in *Bild* as saying "people have nothing to be concerned about!" Dr. Mitton was also reported in an English science magazine as suggesting that all such "alarmist theories" were based on "myth and magic" and should be dismissed. The scientific community at large shares this opinion.

I wish that I could also share these sentiments, but I am not convinced Dr. Mitton is correct in her assessment of the situation. The critics and debunkers who are so keen to point out that planets are too far away as to exert any significant gravitational pull on the Earth, even though the entire solar system is held together by gravity, could be missing a vital key element in the equation—electromagnetism. Of the four fundamental forces described by physics, gravity is weaker than electromagnetism, and it is the electromagnetic fields of the solar system bodies that are linked together.

Therefore a planetary alignment is in essence also an

alignment of the planetary electromagnetic fields, which could directly disrupt Earth's atmospheric conditions and tug at the solar surface, generating powerful solar storms which in turn affect the Earth.

Double Whammy in May: Planetary Alignment with Sunspot Maxima

Coinciding with the planetary alignment on May 5, 2000, will be the peak of the eleven-year sunspot cycle, or *sunspot maxima*, which will most definitely strengthen the possibility of powerful earthquake activity around the month of May. Even in the months of March and April, at the peak of sunspot activity and leading up to the planetary alignment, strong seismic activity could already be felt in sensitive areas around the world.

The current sunspot cycle, anticipated to attain maximum amplitude toward the end of 1999 and up to the first half of the year 2000, will be above average compared to sunspot maxima in other years. The influence of high sunspot activity coupled with an alignment of six planets closely grouped could be really explosive.

Almost twenty years ago, Russian scientists concluded that there was a direct link between solar activity and earthquakes. They observed that powerful solar storms disturb the Earth's magnetic field by as much as a thousandfold, causing the molten interior of the planet to move in powerful circular currents. These massive currents place enormous stresses on the interplay of the tectonic plates, resulting in earthquakes and volcanic eruptions. They also observed that as the Sun approached the peak of its eleven-year sunspot cycle, earthquake and volcanic activity was greatly heightened due to greater stresses on Earth's molten core.

Drs. Gribbin and Plagemann also mention in *The Jupiter Effect* that when sunspot maxima peaked, a marked increase in seismic activity occurred. Conversely, seismic events decreased at sunspot minima, when least sunspot activity occurred.

According to my knowledge and understanding of the planetary lineup, the breeding ground is ripe for one of the most catastrophic earthquakes recorded this century to rock the planet sometime in May 2000.

Prior to predicting dates for Californian earthquake activity several years ago, it was necessary for me to study previous earthquake data for that region. From earthquake data obtained from USGS (United States Geological Survey) dating back some 200 years, I observed that the larger Californian earthquakes—including the great San Francisco quake of 1906—seemed to occur when Jupiter was configured, especially by Mars in either heliocentric (Sun-centered) or geocentric (Earth-centered) positions.

The "Jupiter Effect:" May 2000

The vital focus and key to the powerful earthquake in the month of May 2000 will be the giant planet Jupiter. This very powerful quake could occur within the first week of May around the time of the planetary alignment, or it could come later in the month. It will most likely take place somewhere between May 4–22. Within that time frame, the most critical periods are May 4–12 and May 17–22.

On May 7–9 the Sun, Mercury, Jupiter, and Saturn are at their closest point together at 18 degrees in the sign of Taurus in geocentric longitude. When they do, they will be triggering another event which occurred nine months earlier, on August 11, 1999.

On August 11, 1999, the planets form a "Grand Cross" in geocentric longitude. This configuration will involve a solar eclipse at 18 degrees Leo, Mars at 17 degrees Scorpio, Saturn at 17 degrees Taurus, and Uranus at 14 degrees Aquarius. When a solar or lunar eclipse occurs at the same time that a major pattern is forming between the planets, severe weather, hurricanes, and seismic activity will ensue.

Exactly why an eclipse degree should be of any significance nine months after the fact is not that simple to explain, but an eclipse point is very important in timing of natural disasters. A solar eclipse's position becomes a "fixed point" in space—rather like a spider's web waiting for some unsuspecting fly to trigger the web. The "fly" in this case happens to be any planet that moves into the web and triggers the eclipse point. Depending on where the other planets are at the time, the effects of a solar eclipse can be immediate or can occur many months later, when the eclipse point is triggered by one or more planets. When this occurs, a natural disaster will strike (either immediately or several days later) in the form of severe weather or seismic activity. The mechanics involved in producing this effect remains a mystery—but it works.

The Martian Trigger

Out of curiosity to find out how many solar eclipses had played a role in major historical disasters, I compiled data showing the paths of the solar eclipses from 79 A.D. (when Mount Vesuvius buried Pompeii) until today. The results were very interesting. More than fifty percent of the time, natural disasters were preceded by a solar eclipse falling over that region days, weeks, or even months prior to the event, though rarely beyond a two-year time frame.

The obvious culprit was the planet Mars, which takes 1.88 years to orbit the Sun, coinciding with the two-year time frame. After that, I began to observe the orbit of Mars moving across the eclipse points, and how it triggered either severe weather or seismic activity. (Other planets do play a part as well, especially the outer planets and fast-moving Mercury.)

When I first started learning about weather forecasting, I noticed that a solar eclipse occurred on June 28 1982, at 29 degrees in the sign of Gemini. The equivalent of 29 degrees of Gemini in geocentric longitude is 89 east longitude, over the Bay of Bengal in India. A year later, towards the end of June, the planet Mars moved across that eclipse point and triggered a tropical cyclone which roared up the Bay of Bengal. The storm raised a storm surge that swept into Bangladesh, drowning thousands of people. Several days before the event a lunar eclipse had occurred at 93 east longitude (also over the Bay of Bengal), and helped to strengthen the earlier solar eclipse point.

Centuries ago in India when the sky suddenly darkened with the shadow of a solar eclipse, people began to pray in fear of what was to befall them. Were they merely steeped in superstition, or did they know from previous experience that sooner or later some natural disaster would befall that region?

(*Editor's note*: On August 17, 1999, as the Moon contacted Mars and triggered the August 11 eclipse point, western Turkey was struck by a massive earthquake that killed thousands, injured tens of thousands, and caused billions of dollars of damage. Turkey was directly in the path of the eclipse.)

Nostradamus' Prediction of a Great Earthquake on May 10–11

> "The Sun in twenty degrees of Taurus there will be a great earthquake.
> The great theater, full, will be ruined.
> Darkness and trouble in the air, sky and land
> When they call upon God and the saints"
> (Quatrains: Century 9; 23)

I do not place too much emphasis on predictions that were written 400 years ago by the French seer Nostradamus (1503–1566). Nostradamus' predictions were cleverly concealed in his complex quatrains and a cause for consternation ever since between the various interpreters of his works. However, it is worth noting that his prediction of a devastating earthquake around the May 10–11 time period does seem to fit the conditions of the May 2000 planetary alignment.

Nostradamus was an astrologer, and obviously knew that every year the Sun approaches 20 degrees of Taurus around May 10–11. The influence of the Sun alone is rather insignificant, unless strengthened by Mars or a grouping of the major planets.

His prediction of an earthquake occurring around May 10–11 is very interesting since no earthquake of such magnitude has ever occurred around this date since Nostradamus made this prediction … yet!

So what do the planets have in store on May 10–11, 2000? First of all, the fixed point of the solar eclipse of August 11, 1999, at 18 degrees of Leo (configured in a Grand Cross by Mars in Scorpio, Saturn in Taurus, and Uranus in Aquarius) will be triggered on May 5, 2000, by the combined influences of the Sun, Moon, Mercury, Venus, Jupiter, and Saturn.

On May 10–11, 2000, not only will the Sun will be at 20 degrees, but also Mercury, Saturn and Jupiter (within 2 degrees). All three planets, conjoined with the Sun, will be making a right angle to the planet Uranus. This is very significant in earthquake prediction, as next to Mars Jupiter and Uranus are the major earthquake players. At the same time the heliocentric (Sun-centered) position of Mercury will also be in Taurus, while the heliocentric position of Mars will be opposing Pluto.

From May 17–22 Mars in geocentric (Earth-centered) longitude will be conjoined by Mercury and opposing Pluto. At the same time, Saturn will be in conjunction with Jupiter and both will making an exact square to Uranus. These influences are extremely powerful and can also trigger a major quake.

May 4–12 and May 17–22 are critical periods for earthquakes and possible volcanic eruptions as well. If nothing transpires in early May, then look to May 17–22—especially May 18—as the next critical period for a major event. And that event could directly affect California.

Another of Nostradamus' prophecies worthy of note describes events predicted to take place in the seventh millennium—the year 2000, during spring (May) and October.

> "There will be omens in spring, and extraordinary changes thereafter, reversals of nations and mighty earthquakes ... And there shall be in the month of October a great movement of the Globe, and it will be such that one will think Earth has lost its natural gravitational movement and be plunged into the abyss of perpetual darkness."
> (Nostradamus' preface in his epistle to Henry II)

This prediction is, in my opinion, a vivid description of a powerful volcanic eruption. The worst two periods for seismic activity in the year 2000, according to my calculations, are in spring (northern hemisphere) and October—unnervingly reminiscent of this quatrain! During the months of May and October, powerful quakes, volcanic eruptions, along with fierce weather, will target the planet.

In many instances prior to a violent volcanic explosion, powerful earthquake swarms can cause subsidence and the displacement of large land areas. To those living in affected areas, it would indeed appear as though the Earth had "lost its natural movement." When southern California was rocked by a series of powerful tremors and aftershocks in April to June of 1992, the ground swayed around so much that people began taking motion sickness tablets.

The phrase "... the Earth will be plunged into the abyss of perpetual darkness ..." is also in keeping with a volcanic ash cloud that can follow a powerful volcanic blast. In many cases, the surrounding countryside can be darkened for days by thick, choking volcanic ash falling from the sky. For example, the biggest recorded volcanic eruption mainland Europe has witnessed came on August 24 in 79 A.D. when Mount Vesuvius, towering 4,000 feet above the Bay of Naples, erupted in a series of violent explosions. It buried the city of Herculaneum under a twenty-foot-deep volcanic tomb of mud and lava, and blanketed Pompeii on the other side of the mountain in a twenty-foot-thick carpet of ash.

According to historical records, all of the 20,000 people who perished had ample time to evacuate the two cities during the daylight hours. The warnings started with swarms of sharp earthquakes that rocked the area, causing major displacement of land around the Bay of Naples and other parts of the Italian coastline.

During the early hours of the morning of August 24, the volcano entered its final stage, convulsing violently in a series of death-dealing explosions and causing more land displacement. The only escape route from Pompeii, now enveloped by death-dealing thick, choking clouds of ash, was by sea. But further subsidence around the docks and rough treacherous seas from convulsing seismic activity made escape impossible.

The Nostradamus quatrains for May and October seem to fit the time frame for major events in the year 2000. The planetary aspects coming into play around these months would certainly indicate a major earthquake in May, followed by a major volcanic eruption in October. Volcanic eruptions have occurred before under the combined influences of Mars, Jupiter, and Pluto, as will be the case in early and late October 2000.

Nostradamus' Prediction on Floods "As the Great Century Draws to a Close"

Another of Nostradamus' predictions could also fit with this time period, if his phrase "as the great century draws to a close" is actually in reference to the twentieth century drawing to a close, and the beginning of the new millennium.

> "Rain, blood, war, famine and disease as the great century draws to a close.
> So many deluges and so many floods that hardly any land will be left untouched by water."
> (Nostradamus' preface in his epistle to Henry II).

The "biblical deluges and mammoth floods" that are likely to occur in late June, July, and August in the year 2000 are certainly consistent with what Nostradamus foresaw at the close of the century. In the last 400 years, there have never been floods of the catastrophic proportions predicted by Nostradamus.

Apart from devastation wrought by floods, and in order of destructive potential, the worst months for the year 2000 are May, October, August, and, to a lesser extent, March and April. At times, in writing weather and seismic forecasts, it has been extremely difficult to figure out exactly how the planetary patterns will affect that particular location. In those instances when I haven't been quite sure which way an event will be weatherwise or seismic, I have given two options. In areas that are not prone to seismic activity, the influence is more likely to be severe weather patterns.

In either case, the most likely regions of the world to be affected in May appear as follows:

Northern California or Turkey/Iran: Massive Quake, May 7–22

There are two regions of the world earmarked for an earthquake of great magnitude resulting from the planetary alignment of May 2000 coupled with high sunspot activity. One is northern California, possibly affecting San Francisco. The other is Iran or Turkey, situated 180 degrees away on the opposite side of the world. Both regions, however, will not be affected simultaneously. It will be one or the other. Location charts pinpoint northern California as being the more likely of the two.

Mexico: Tremors, May 8–14 and May 17–24

Mexico City and points east, such as San Luis Potosi and Tampico, could experience powerful tremors between May 8–14 or May 17–24. There is also a possibility of volcanic activity from Popocatapetl near Mexico City at this time.

Hawaii: Volcanic Activity, May 14–24

The big island of Hawaii may experience some volcanic activity from Kilauea around May 14–24.

Montserrat And Martinique (Mount Pelee): Volcanic Activity, May 2–22

Perhaps one of the most volatile regions of the world for dangerous explosive volcanic eruptions is the Lesser Antilles group of islands, strung out like green emeralds in the West Indies. Here the volcanoes are very unpredictable slow boilers, simmering away like natural time bombs, and ready to explode at any time.

The tiny island of Montserrat is located to the north in the Leeward Island group. In recent years, Montseratt's volcano has been highly active with frequent eruptions showering ash over Plymouth, the capital city. So far, though, the Soufriere Hills volcano has not erupted on a grand scale.

Further south in the Windward island group are the islands of St. Vincent and Martinique, where two other volcanoes are located. Although St. Vincent's La Soufriere volcanic eruptions can be spectacular, they are not violent like the explosive eruptions of the Mount Pelee volcano on Martinique, which is considered to be the most dangerous volcano in the West Indies.

On May 7 and 8, 1902, these two dangerous volcanoes erupted in unison, killing 30,000 people on Martinique and 2,000 on St. Vincent. The La Soufriere volcano had been rumbling away since early February 1902, but Mount Pelee only began to show signs of life when it began to rumble in early April, one month before it exploded.

The powerful eruptions of Mount Pelee and La Soufriere coincided with the solar eclipse of May 7, 1902, which directly affected that region as shown on longitude and latitude charts. The solar eclipse at 16 degrees of Taurus was exactly ninety degrees (square) to the planet Jupiter at 15 degrees of Aquarius. Mars and Mercury were also in the sign of Taurus, close to the eclipse point and close to making a ninety-degree angle to Jupiter as well.

From May 2–22, 2000, the planets will once again pass over Mount Pelee, La Soufriere and Monseratt's volcano—and will form a similar pattern to that of May 7, 1902. This time however, Jupiter (the earthquake planet), Saturn, Mercury, Venus, and the Sun will be in the sign of Taurus and form a ninety-degree angle to Uranus (another earthquake planet) in Aquarius.

The combined influences of all these planets could well trigger another major volcanic eruption in that region. Warning signs of impending volcanic activity could begin around March and April. If this is so, May could bring a major volcanic eruption in this chain of islands.

Indonesia: Fierce Storms, May 17–24

The same scenario applies elsewhere in the world, especially for the Indonesian volcanoes on the islands of Sumatra and Borneo, and the Rabual volcano in New Britain, Papua New Guinea. The Lesser Sunda Island group could be targeted by powerful, destructive storms around May 17–24 which could move south into the Arafura Sea and possibly form a tropical depression.

Northeast Argentina: Tremor, May 8–17

Northeast Argentina could experience a powerful tremor between May 8–17.

Fiji: Tremors or Fierce Storms, May 12–22

In the South Pacific Ocean, the island of Fiji could be affected by either fierce storms resulting from a very late tropical cyclone around May 12–22, or offshore earthquake activity.

Philippines, Southeast China, or Taiwan: Tremors, May 7–12

The northern section of the Philippine island of Luzon could sustain a strong tremor from May 7–12, or possibly an offshore quake. Powerful tremors could also shake regions around southeast China and Taiwan at the same time.

India: Fierce Storms, May 10–13 and May 18–24

Fierce storms and possible tornado activity will move from west to east across India around May 10–13. Around the same time, a cyclone could form in the Bay of Bengal and move toward the central Indian coastline. Conditions between May 18–24 indicate that central India could be hit by a major quake or a very destructive storm.

U.S.A.: Fierce Storms and Tornados, March, April, and May

Major floods may also be experienced in various parts of the world from April and into May as well. For the United States, the months of March and especially April and May will bring wild storms with fierce destructive winds and powerful tornado outbreaks. Some of the affected areas will be the Dakotas, the Midwest, Texas, the Deep South, the greater Ohio Valley, and the northeastern states.

Dates indicated for powerful tropical storms are around May 10–22 and April 13–23. These influences are too early in the season to cause northeast Pacific and Atlantic hurricanes, but could trigger late tropical cyclones in the southwest Pacific, the Coral Sea, or the Indian Ocean.

July Eclipses Spawn Major Deluges

In the months of April and May there may be severe flooding in various locations around the world, but these inundations will be nothing compared to the floods coming in July. Heavy rainfall may begin by mid-June, already saturating the ground before the month of July arrives when three eclipses—two solar and one lunar—will bring deluge after deluge, resulting in floods of near-biblical proportions.

A solar eclipse is when the Moon passes between the Earth and the Sun, effectively blotting out the Sun. When the Earth passes through the main shadow cone, or *umbra*, the eclipse is "total." In annular or partial eclipses, portions of the Sun are still visible.

A lunar eclipse occurs when the Earth is between the Sun and Moon. There are three kinds of lunar eclipse: total, partial, and penumbral.

Solar and lunar eclipses follow each other by two weeks, and occur twice yearly six months apart.

The influence of a total solar eclipse is generally considered to be far greater than that of a partial solar (or lunar) eclipse. At the time of a total solar eclipse the solar wind is deflected by the Moon, causing radical changes in electrical Earth current and the geomagnetic fields.

To have three eclipses in a row is a relatively frequent occurrence. The influence of an eclipse however, really depends on where the other planets are during the event. If several planets are grouped near the eclipse point, as they will be in July 2000, the effects will be more severe. This is true despite the fact that neither of July's solar eclipses will be total eclipses.

Geocentric longitudes in zodiacal equivalent are referred to as "geodetics," which can often be very helpful in weather forecasting, but only if and when the geodetic points also match location charts drawn up for the same longitude and latitude. If there is no match, then the longitude and latitude charts always take priority over geodetic points, for these "location" charts determines where the influence of the planets will be most pronounced.

In this case, the first solar eclipse, taking place on July 1, 2000, will be at ten degrees in the sign of Cancer. On a world map, ten degrees of Cancer is equivalent to 100E longitude, which falls over China. However, location charts around the world show the eclipse falling over northwest India, so although China may be influenced to a degree, it is India that will receive the full force of the solar eclipse.

The major area of concern in India is in the northwest section of the country near Rajisthan and Gujuvat (and possibly neighboring Pakistan), which will collect the brunt of severe monsoon floods from mid-June to end of July. East India will also be affected by flooding from July 8–22, but this region should not be as severely affected as the northwestern states.

This is the time of year when the monsoon rains begin to fall over India. To a greater or lesser extent, deluges and flooding are a common occurrence during this season. Every year in India, many lives are lost through monsoon flooding.

For example, in the summer of 1978, the most severe monsoon rains in living memory raged throughout northern India. Official reports had the deluge dislocating millions of people, destroying more than 600,000 homes, and sweeping 1,000 people to their deaths. Unofficial counts put the figure much higher, more in the vicinity of 4,000 dead. The neighboring country of low-lying Bangladesh didn't fare much better when severe monsoon rains flooded 20,000 square miles of the country and drowned 2,500 people in August 1974. Monsoons as extreme as this seldom happen on such a large scale.

However, the monsoon period for India in July 2000 will be extremely severe, possibly the worst in decades, or even in this century.

Venus, Mars and Mercury Retrograding in Cancer: Horrendous Floods

Through hundreds of years of observations, the ancient astronomers noted that great deluges and floods occurred when several planets transited together through the constellation of Cancer the Crab. This is especially true of the fast-moving planets Mercury, Venus, and Mars. This effect is even more pronounced if one of the planets happens to be moving in "retrograde" motion at the same time.

Planets, of course, do not really move in retrograde motion (i.e., backwards), for in reality they are constantly orbiting the Sun. However, the planets appear to move backward relative to the Earth, seeming to slow down to a stationary position then moving in the opposite direction. Eventually, the planets then begin to slow down to a stationary position before moving forward again.

Retrograde movement is like the spokes of a fast-moving wheel that appear to move backward. It's only an optical illusion, but in principle, it demonstrates the same type of effect as retrograde motion.

When planets are at their stationary points, either about to move forward or retrograde, their influence on weather patterns is far more powerful than normal. Planets at their stationary positions greatly disrupt and disturb the atmospheric conditions around the Earth, generating destructive storms with gale-to hurricane-force winds that rage across the planet.

Planets moving in retrograde motion—especially the faster-moving inner planets—can produce some mighty fierce storms as well. In fact, storms that form under retrograde planets tend to be far more violent than usual.

Fierce category five hurricanes like Gilbert and Hugo formed under the effects of retrograde planets. Every year, most of the planets move in retrograde motion for shorter or longer periods.

So what effect will a grouping of retrograde planets in the sign of Cancer have on Earth's weather patterns from mid-June, July, and early August?

From June 21 until mid-July 2000, the Sun, Mercury, Venus, and Mars will all be gathered in the sign of Cancer. Combinations of these planets in any sign have always brought about severe weather conditions, deluges, and floods to wherever their influence is greatest in the world charts.

It takes only one of these planets in Cancer to increase rainfall over a certain area, but if two or more are gathered together in that sign you can expect major floods to follow. If one of these planets happens to be retrograde at the same time, the influence is further intensified.

So from mid-June until mid-July, when the Sun, Venus, Mars, and retrograde Mercury are all in the sign of Cancer, flooding on a massive scale may occur. On July 1, when the solar eclipse takes place in Cancer at the same degree as Mars, and with Venus and retrograde Mercury also in Cancer, the deluge will really begin to intensify.

Unfortunately, this is not the end of the rain, nor the floods. The great deluge will last from mid-June throughout July, and continue into early August. The floods anticipated for June, July, and early August will be stronger than those in April and May. Hurricanes and typhoons forming in the months of June, July, and early August will bring devastating deluges and catastrophic floods should these storms move ashore anywhere around the world.

On July 16, when the lunar eclipse occurs, the Sun will still be in the sign of Cancer, as will Mars and Mercury (now moving forward), indicating that the rains will continue until the end of July. The rain during this period will not be as severe as in the first two weeks of July. Nevertheless, continued rainfall will further exacerbate an already critical situation. Between June 28 and July 16, powerful hurricanes and earthquake activity could also be experienced around the world.

The flip side to flooding is, of course, drought. In the year 2000 there will be countries that will experience severe drought conditions, while others will experience severe flooding. The reversal of climatic conditions not only affects opposing hemispheres, but occurs within the same country as well. In the United States, for example, the northern states can be undergoing severe cold, wintery storms, and deluges, while the southern states experience higher temperatures and drier-than-normal conditions.

Eastern U.S.A., Including Mississippi and Ohio Rivers: Floods, June 26–July 16

In the United States, heavy rainfall and flooding can be expected for the Eastern states from June 26–July 16. The Mississippi and Ohio rivers could well overflow their banks, while many of the Midwestern states may also receive heavier-than-normal rainfall.

West Coast, U.S.A.: Floods and Mudslides, July 8–31

The coasts of California and perhaps Oregon will also receive several powerful storm systems that will most likely trigger floods and mudslides around July 8–18 and July 22–31.

Southern Mexico and Texas: Tropical Storms Cause Major Flooding, July and August

Several tropical storms and/or hurricanes will form in the Pacific off southern Mexico in July and August. Several of these systems will move toward the Mexican coastline. These tropical storms/hurricanes will cause major flooding to southern parts of Mexico. They will possibly move in as a rain band northeast into Texas, dumping a deluge there as well, especially in July.

Northwest Pacific Ocean and/or the Philippines: Fierce Typhoons, July 1–16

In the northwest Pacific Ocean from July 1–16, a powerful typhoon may form east of the Philippines and move toward Samar or southern Mindanao. If it moves west toward the Philippine islands, a storm surge could inundate the coastline. Alternatively, this influence in the northwest Pacific could trigger a seaquake and raise a powerful tsunami which may affect Mindanao.

Western African States and Northern Italy: Fierce Storms and Floods, July 1–16

Across the Atlantic Ocean from the United States lies the West African coastline, where powerful storm systems often bring much-needed rains to the West African states.

From July 1–16, fierce storms off the coast of West Africa will sweep across Mauritania and Senegal and then move northeast. Major flooding will occur wherever this system travels.

Location charts for West Africa and Europe indicate that either this system or a new one will move across northwest Africa, then sweep up over the Mediterranean into northwest Italy, where it will bring major floods. From July 23–August 12 there will be several powerful storm systems that will affect northwest Italy.

Northwest Pacific Ocean and East Turkey: Earthquakes, July 1–16

Apart from triggering major floods around this time period, the powerful influences from the solar eclipse could also trigger major quakes, particularly in the northwest Pacific Ocean and eastern Turkey.

Ireland: Floods, July 8–19

From July 8–19, fierce storm systems forming in the north Atlantic could move across Ireland. If this should happen, heavy deluges could cause floods.

Ireland and Southwest England: Fierce Storms and Floods, July 31–August 20

Another round of storms, even more severe than the last, are likely to form off the west coast of Ireland around July 31–August 20 and move across Ireland into southwest England. If this occurs, gale-to hurricane-force winds could be whipped up.

The Neptune Connection

Long before the planet Neptune was discovered in September 1846, the Romans had already given their mythical god of the sea the name "King Neptune." The most bizarre twist to this little tale, and pertinent to weather forecasting, is the fact that the planet Neptune nearly always appears in a predominant position in location charts prior to major flooding in that particular region of the world. The oddest thing is that the Romans knew about Neptune's influence long before the ancient astronomers knew the planet existed.

Not only does Neptune somehow appear to be connected with floods, but shows up in storm surges and tsunamis as well. Although the planet Saturn can also indicate major floods, Neptune's influence is more predominant. There is no apparent rhyme or reason why the planet Neptune, the eighth planet from the Sun, should influence large bodies of water, but it most certainly does.

The final solar eclipse on July 31 is in the sign of Leo, where it is opposing (180 degrees). Neptune in the sign of Aquarius. Between August 6–14, Mercury and Mars move together over the July 31 solar eclipse point while opposing the planet Neptune. Not only will this bring another round of deluges and severe flooding, but fierce weather, powerful hurricanes, typhoons, earthquakes, and volcanic activity can be expected.

New Zealand and Indonesia: Earthquakes and Eruptions, July 27–August 12

The North Island of New Zealand will be affected by this fierce influence which ever way it swings. In Indonesia, southern Sumatra or western Java may also experience seismic activity and volcanic eruptions.

Papua New Guinea, New Britain, Bougainville: Storms or Tremors, August 5–16

August 5–16 could see either fierce storms or powerful tremors along the Bismarck Archipeligo. There is a strong possibility that Rabaul's volcano may also be activated during this time.

New Zealand and Iran: Storms or Tremors, August 22–September 3

Although indications are for destructive storms with gale-force winds for the North and especially the South Island of New Zealand, there is also a very good chance that powerful tremors could rock the country. At the same time, major tremors or severe storms are also indicated for Teheran, Iran.

Southern Italy: Tremors or Eruptions, September 5–17

Seismic activity, either in the form of strong tremors or possible volcanic activity, is indicated for southern Italy at this time.

Widespread Cataclysms in October

Major volcanic activity, powerful earthquakes, hurricanes, and severe weather will wreak havoc in October due to one of the most powerful planetary configurations of the year. This will bring fierce hurricanes, typhoons, and storms with very destructive winds and hail, but more important is that around this time period a volcano is likely to explode in a full scale eruption near any of the locations pinpointed. Powerful earthquakes can also be anticipated around this time period. The areas most affected are: in Mexico, Oaxaca; in Japan, Honshu and possibly Tokyo; in the Philippines, Mindanao; in the U.S.A., Oregon and northern California; and all of southern Italy.

The Italian volcanic chain should be very closely watched around this time, especially Mount Etna and the unpredictable Mount Vesuvius (which is currently dormant). Throughout history giant Mount Etna, on the island of Sicily, has claimed over one million lives. In July 1998 Mount Etna began a new eruptive cycle, and could have a full-scale eruption at any time. If either volcano starts showing signs of life in 2000, watch out for October.

Powerful earthquake activity could also be experienced in southern Italy around this time, as well as in the Mexican state of Oaxaca. Southeast Honshu and Tokyo could experience powerful tremors as well as a powerful typhoon strike.

The Philippine island of Mindanao, along with Oregon and northern California, are likely to be targeted by extremely dangerous storms, although the possibility of very powerful tremors should also not be discounted.

The Most Destructive Storms of 2000

While all of the storm systems listed chronologically below will be intense and spawn powerful winds, some of them will be more destructive electrical storms, with fierce lightning, gale force winds, hail, and in some cases, tornados. The following storm list has been rated accordingly from the most intense to the least intense storms.

If *any* of the dates of the listed storms correspond with storms brewing in your local area, be aware that the storm has the potential to be extremely dangerous. During the whole of July, August, and October, storms are likely to be particularly severe

Period	Intensity
1/14–21	Severe
1/28–31	Strong
2/11–18	Strong
2/27–3/5	Strong
3/15–28	Violent, destructive
4/5–22	Violent, destructive
5/4–14	Violent, destructive
5/17–24	Extremely violent and destructive
5/26–6/4	Severe
6/16–30	Severe
7/1–16	Violent, destructive
7/31–8/16	Extremely violent and destructive
8/20–9/3	Extremely violent and destructive
9/10–23	Severe
10/1–10	Extremely violent and destructive
10/13–20	Violent, destructive
10/27–11/3	Violent, destructive
11/13–17	Strong
11/19–23	Violent, destructive
12/17–24	Severe
12/28–31	Severe

CHAPTER THREE

Tropical Cyclones, Hurricanes, and Typhoons in 2000

Powerful tropical storms, considered to be the greatest storms on Earth, originate over warm tropical waters, which are essentially "heat engines" with a central core that has a higher temperature than the surrounding atmosphere. Combine warm ocean temperatures with warm, moist, unstable atmospheric conditions, and tropical storms have a suitable breeding ground in which to spawn. Although tropical storms are named differently in different parts of the world, they are basically the same type of storm.

There are three basic stages that tropical storms pass through when they are developing, and also when they are dissipating. Initially forming as a tropical low or tropical depression, these systems intensify into tropical storms, then one step further, to become mature hurricanes, cyclones, or typhoons.

These great storms are categorized from weak (category one), where winds range between 74–95 mph (125 kph), to catastrophic (category five), where winds can clock up to 156 mph (280 kph), or even higher, with devastating results when making landfall. Fully matured systems possess massive physical structures that extend vertically from the surface of the sea to the base of the stratosphere. They cover enormous geographical expanses of land and sea, usually some 100–400 miles in diameter.

These huge whirlwinds of air, moving in a very tight spiral around a relatively calm center or "eye" of extremely low pressure, have life spans lasting from days to weeks, and can trace a path several thousand miles long across the ocean. The lower the barometric pressure, the more intense the hurricane becomes.

In the United States, hurricanes target the coastline either on their "wet" side, bringing deluges, or their "dry" side, resulting in tornado swarms.

Fierce hurricanes and their close relatives are greatly feared and the most costly of all natural disasters. On making landfall, these feared storms can affect miles of coastline, flatten everything in their path, cause extensive flooding, generate tornado swarms, or whip up sea waves, sending storm surges crashing onto shorelines.

A severe category five system can raise a storm surge eighteen feet or higher, depending on the storm's intensity. Powered by the fierce winds, the sea waves are pushed ahead of the storm onto beaches and into estuaries. These powerful waves crash repeatedly upon coastlines, sweeping them clean of all life and vegetation.

Many of the world's coastlines are vulnerable to hurricane-driven storm surges. The more the population swells in our coastal towns and cities, the more costly these great storms of destruction become, and the greater the toll of human life.

Of course, where you live makes a difference. On the coast of the Bay of Bengal in India or Bangladesh, for example, the death toll from storm surges is very high when compared to the United States. This is primarily because in the richer nations emergency services normally order the population to vacate coastal regions well in advance of a hurricane threat.

The months of August and October in the year 2000 will be the most critical months for severe, destructive hurricanes and typhoons. And although September's storms may also be quite powerful, they will not be as severe as August. Should any of August's hurricanes or typhoons move towards shorelines anywhere in the world, be aware that the nature of these storms will be extremely dangerous, with fierce, destructive winds.

All tropical storm forecasts are dependent on the southern oscillation index and El Niño development. El Niños retard tropical storm development, and if by chance one happens to develop in the year 2000, many of the tropical storms listed below will not develop.

Sometimes a tropical storm will form and make landfall within several days, while at other times a tropical storm system can last up to three weeks in the open ocean. Therefore, the following dates relate to the formation of tropical storms and not for their life spans. At times, a tropical storm will form and make landfall within the given time frame, but this is more the exception than the rule.

"Wild Card" Factors in Forecasting

With so many overpopulated "hotspots" around the world periliously on the brink of becoming cataclysmic natural disasters, the coming century has the potential for being one of the worst in the history books for resultant loss of life due to natural disasters.

Complicating matters further is the climatic factor of global warming, which, by all appearances, could become even more severe in years to come and even create a sort of permanent El Niño. If the effects of the mini-ice age that began 600 years ago finally disappear altogether, global warming could accelerate much faster than it is today.

An example of just how much the world's oceans have been heating up over the past twenty years is shown in the heating trends of El Niño verses the cooling trends of La Niña. Over the hundred-year period from 1882-1982, there have been a roughly equal number of El Niños and La Niñas. However, in the past twenty years there have been only three "cool" La Niñas, compared to seven "warm" El Niños. Since 1982, the number of El Niños has more than doubled.

On the other hand, a cooling phase could be just around the corner. The truth is, with all our technology and simulated climate models, no one has got the whole picture, or really knows of what nature has in store for us.

No one really knows either, exactly what the impact of the great planetary alignment of May, 2000 will be on the

Earth. Although geological evidence dating back to 4000 B.C. (when a similar planetary alignment took place) indicates that earthquakes and floods raged on the planet then, are "rock records" all that reliable?

Every single weather forecast or earthquake prediction that I have made throughout the years has been based either on my own observations or on documented events; in other words, a database to draw conclusions from. The May 2000 planetary alignment, however, where six planets will be lining up almost in a straight row, is unprecedented. There is no database to draw conclusions from, so I can only guess how severe the impact will be.

I very much doubt that the planetary alignment will pass without incident because of the higher-than-normal sunspot activity, which enhances seismic activity, and electromagnetic stresses which at times have triggered earthquakes.

Storm Schedules for 2000

In the following tables, storm forecasts extending from January to December 2000 have been compiled for almost every country, although only the major influences have been listed, and not every storm that may come along. These forecasts are very generalized and may target an area on the same longitude line, but a latitude north or south of the specified area. The same rule applies to earthquake activity, which may also take place in latitudes further north or south of the specified location.

Severe storms systems will operate differently in different climates, as well as in different months. During hot steamy summer months a storm system will be far more severe, with a strong likelihood of hail, than if the summer is cool. In winter months, a severe storm system is more likely to dump piles of snow, blow icy gales, and in some cases bring dreaded ice storms, as opposed to deluges, which would be the result if it were summertime.

Weather patterns, generally speaking, move from west to east—either due east, northeast, or southeast. If a predicted powerful storm hits west of your location, your city may well be next in line for thunderstorms.

In regard to the date ranges predicted for fierce storm systems, tropical cyclones, or hurricanes, please take note of the prevailing weather patterns shown on television at that time. If the patterns fit the long-range forecast, then and only then, take precautions. Major storms will generally hit only once within the specified time frame and not every single day. Once the storm hits, the influence—and the forecast—is finished.

Tropical Cyclones
SW Pacific, Coral Sea, Arafura Sea, Timor Sea, Indian Ocean, Arabian Sea, Bay of Bengal

Date	Location	Hot Spot	Intensity
1/17–25	Coral Sea		Severe
2/5–8	SW Pacific	New Caledonia	
2/27–3/4	Arafura Sea		
2/28–3/4	Coral Sea		
3/20–28	Coral Sea		Severe
3/20–28	Timor Sea	Melville Island	Severe
4/13–22	Indian Ocean	off W. Australia	Severe
5/10–14	Coral Sea		

Date	Location	Hot Spot	Intensity
5/10–22	Indian Ocean		Severe
5/12–22	Bay Of Bengal		
5/12–22	South Pacific	near Fiji	Severe
7/1–16	Indian Ocean		Severe
7/8–18	Bay Of Bengal		Severe
8/ 23–31	Arabian Sea	off Pakistan	Severe
9/13–20	Indian Ocean	off Somalia	
10/2–10	Indian Ocean	off Madagascar	Severe
12/18–24	Gulf of Carpentaria		
12/18–24	South Pacific	East of Vanuatu	

Typhoons
NW Pacific, E. China Sea, S. China Sea, Andaman Sea, Philippine Sea

Date	Location	Hot Spot	Intensity
4/11–22	S. China Sea		
5/17–30	NW Pacific		
6/20–30	NW Pacific.		
7/1–16	NW Pacific		Severe
8/15–9/2	NW Pacific		Severe
8/15–31	Andaman Sea	North of Sumatra	Severe
8/22–31	NW Pacific	off Japan	Severe
9/15–23	NW Pacific	off Japan	Severe
9/17–23	Philippine Sea	off N. Philippines	Severe
10/1–10	NW Pacific	off Taiwan and/or Japan	Severe
10/14–24	NW Pacific	off Japan	Severe

Hurricanes
NE Pacific and N. Central Pacific

Date	Location	Hot Spot	Intensity
6/20–30	NE Pacific	off Mexico	
7/1–15	NE Pacific	off Mexico	Severe
7/31–8/12	NE Pacific	off Baja	Severe
8/5–13	NE Pacific	off Mexico	Severe
8/5–13	N Cent. Pacific	off Hawaii	Severe
8/15–24	NE Pacific	off Mexico	Severe
8/22–30	N Cent. Pacific	off Hawaii	Severe
8/24–9/3	NE Pacific	off Mexico	Severe
9/10–23	N Cent. Pacific		
10/2–10	NE Pacific	off Mexico	Severe
10/5–12	NE Pacific	off Baja	Severe

Hurricanes
Atlantic, Caribbean Sea, Gulf of Mexico

Date	Location	Hot Spot	Intensity
6/26–7/7	Gulf of Mexico		
7/8–20	Atlantic		
7/31–8/10	Atlantic		
8/16–24	W Caribbean	Gulf of Mexico	
8/22–9/3	W. Caribbean	Yucatan, Honduras	
8/22–9/3	Atlantic	Florida, S. Carolina	Severe
8/26–9/3	Atlantic	Lesser Antilles	Severe
9/10–23	Atlantic Ocean	North Carolina	
9/15–24	W. Caribbean		
9/16–23	Gulf of Mexico	Bay of Campeche	
10/2–10	Gulf of Mexico	Louisiana	Severe

Chapter Four

Event Timetable: 2000

Europe

Dates	Locations	Events
1/6–12	North Sweden	Storms, deep freeze, heavy snowfall
1/14–20	Holland	Severe storms and cold, deluges, or heavy snowfall
1/14–20	S. Finland, E. Poland, Baltic states, NW Russia, Ukraine	Fierce storms, blizzards, heavy snowfall
1/17–23	N. Central Italy	Fierce storms, deluges, floods, snow
1/20–26	W. Ireland	Torrential rain, floods
1/24–31	Poland	Storms, snow
1/27–31	NW Germany	Storms, snow

Dates	Locations	Events
1/27–31	SE England, Belgium, NW France	Strong storms
1/28–31	W. Ireland	Strong storms
1/30–2/2	S. Italy	Torrential rain, floods
2/ 5–8	W. Ireland	Powerful storms
2/5–8	Holland, S. Finland	Powerful storms, snow
2/5–8	Ukraine, N. Greece	Powerful storms
2/12–16	Holland	Powerful storms
2/17–18	Spain	Storms and floods
2/17–25	S. Italy	Powerful tremors or storms
2/19–23	Hungary, Czechoslovakia	Storms
2/19–23	NW Italy	Storms
2/20–28	Ireland, SW England	Strong storms
2/23–3/3	NW Germany	Powerful storms
3/9–15	N. Italy	Storms, torrential rain
3/11–15	France	Storms, torrential rain
3/11–20	North Sea, Norway	Powerful storms
3/12–22	N. Italy	Torrential rain, floods
3/13–18	Greece	Torrential rain
3/16–22	Greece, Bulgaria, Romania, Poland	Powerful storms and/or tremors
3/16–26	Germany	Severe storms, powerful winds

Dates	Locations	Events
3/20–31	North Sea, France, Belgium, Holland	Fierce storms, powerful winds, deluges
3/22–26	S. Italy	Possible tremors
3/28–4/2	Ireland, England, Germany	Powerful storm systems
3/30–4/7	S. Italy	Tremors, volcanic activity
4/3–6	Greece	Torrential rain and/or tremors
4/6–14	Germany	Powerful storms, floods
4/11–18	North Sea, Norway, Holland, Belgium	Severe storm systems
4/15–5/22	S. Italy	Powerful tremors or volcanic activity
4/20–24	Greece	Storms, strong winds
4/30–5/5	S. France, NE Spain	Torrential rain, floods
5/3–6	SW England	Storms, torrential rain
5/4–12	Holland, Germany, Greece, Bulgaria	Storms, torrential rains
5/17–22	Aegean Sea	Destructive storms or powerful seismic activity
5/17–22	Ireland, W. England	Storms, could be very severe
5/23–25	N. Italy	Storms, torrential rain
5/26–6/2	Holland	Powerful storms
5/27–6/3	Yugoslavia, Greece, E. Poland	Storms
5/28–6/3	Italy	Storms

Dates	Locations	Events
6/2–14	France, Holland, Germany	Storms, torrential rain
6/6–18	N. Italy	Storms, deluges
6/9–14	England	Torrential rain, possible storms
6/16–30	N. Finland, Lapland, Baltic states, NW Russia	Storms, deluges, major floods
6/16–30	Ireland, W.England, Portugal, Spain	Torrential rain
6/16–30	Greece, Turkey	Torrential rain, floods
7/1–10	Finland, Lapland, Baltic states	Storms, deluges, floods
7/1–16	W. Africa, NW Italy	Severe storms, great floods
7/8–19	Ireland, SW England, NW France, possibly Portugal and W. Spain	Powerful storms, major floods
7/23–31	Ireland, S. England	Rain, floods intensify
7/23–31	N. Italy	Storms, deluges, floods
7/29–8/5	S. Italy	Storms, deluges, floods
7/31–8/15	Albania, Greece	Fierce, destructive storms, possible floods
7/31–8/15	W. Ireland SW England	Extremely severe storms, gale-force winds, deluges, and major floods
8/4–14	Baltic states, W. Russia	Fierce storms, deluges, major floods
8/5–12	NW Italy	Fierce storms, deluges, major floods

Dates	Locations	Events
8/15–20	Ireland	Fierce storms, possible gale-force winds
8/22–31	Ireland, SW England, NW France, Spain	Fierce storms, gale-force winds
8/25–31	Germany	Fierce storms, gale-force winds
8/26–9/5	NW Italy	Major earthquake or fierce storms
8/26–31	Aegean Sea	Strong tremors or fierce storms
8/29–9/1	NW Germany	Fierce storms, gale-force winds
9/5–14	S. Italy	Volcanic activity or powerful tremors
9/13–24	E. France	Fierce storms, floods
9/14–23	Czechoslovakia, Hungary, Poland, Yugoslavia	Fierce storms, powerful winds
9/17–25	Italy	Fierce storms or powerful tremors
9/29–10/4	SE England, France	Storms, torrential rain
10/1–10	S. Italy	Major seismic activity, volcanic eruptions, or severe storms
10/1–14	Czechoslovakia, Hungary, Poland	Fierce destructive storms
10/9–12	W. Ireland	Powerful storms
10/13–20	SW Sweden	Powerful storms
10/13–24	Ireland, SW England	Fierce storms, strong winds
10/14–22	S.Italy	Storms
10/15–24	N. Italy, Austria, Czechoslovakia, Poland	Powerful, destructive storms

Dates	Locations	Events
10/20–22	Bulgaria, Romania, Ukraine, W. Russia	Storms
10/20–26	Finland	Strong storms
10/23–25	Holland, Belgium, France	Storms
10/25–11/2	Greece, Aegean Sea, SW Turkey	Powerful storms and/or tremors
10/25–11/3	Bulgaria, Romania, Ukraine, W. Russia	Powerful storms
10/27–31	Portugal, Spain, England	Storms, could be powerful
10/28–11/5	Poland, Baltic states	Fierce storms
11/2–10	Romania, Bulgaria	Fierce storms
11/5–15	Greece, Aegean Sea	Fierce storms or powerful tremors
11/6–11	Portugal, Spain, Ireland, W. England	Powerful storms
11/11–16	Czechoslovakia, S. Poland	Storms
11/12–21	England	Storms, torrential rain
11/13–17	Portugal, Spain, England	Powerful storms
11/18–20	SE Italy	Torrential rain
11/18–25	Greece	Storms
11/19–23	England, France	Powerful storms
11/22–28	North Sea, Holland, E. Germany	Storms, could be powerful

Dates	Locations	Events
12/1–4	SE Italy	Deluges
12/1–8	English Channel, France	Storms
12/5–9	Germany	Storms
12/18–22	Ukraine, W. Russia	Storms, very cold, snow or deluges
12/18–25	Ireland, England	Storms, fierce winds, snow
12/24–31	Central Italy	Powerful storms, deluges, snow
12/25–31	S. Italy	Storms and/or seismic activity
12/27–31	Ukraine	Powerful storms, deluges/snow,
12/27–31	N. Italy	Powerful storms, snow
12/28–31	Ireland, England, Germany	Storms, deluges, heavy snowfall

W. Russia: Latitudes 26E–60E

Dates	Locations	Events
1/14–22	Finland, Baltic states, NW Russia	Deep freeze, strong blizzards, heavy snow
1/21–30	W. Siberian plain	Deep freeze, blizzards, heavy snow
2/5–8	SW Siberian plain	Storms, deep freeze, heavy snow
2/5–11	Finland, Baltic states, NW Russia	Deep freeze, blizzards, heavy snow
2/12–20	NW Russia	Storms
2/19–24	E. Ukraine, W. Russia	Storms, torrential rain
2/27–3/3	E. Ukraine W. Russia	Storms, torrential rain

Dates	Locations	Events
3/8–20	W. Russia	Storms, torrential rain, floods
3/13–20	SE Black Sea, Turkmenistan	Torrential rain
3/20–25	Moscow, SW Russia	Storms, torrential rain
3/28–30	N. Aral Sea	Storms
4/4–10	Ukraine, W. Russia Black Sea	Storms, torrential rain, floods
4/16–22	Moscow, SW Russia	Powerful storms and winds
4/26–5/1	NW Russia, Petrozavodsk	Powerful storms and winds
5/10–18	Kazakhstan	Storms, deluges
6/16–30	NW Russia, Baltic states	Powerful storms, deluges, floods
6/31–7/16	W Russia, Baltic states, E. Ukraine	Severe storms, deluges, major floods
7/16–30	E. Siberia	Powerful storms, floods
8/6–14	Finland, Baltic states, NW Russia	Fierce storms, gale-force winds, deluges and widespread floods
8/20–28	W. Russia	Severe storms, gale-force winds
8/22–30	Uzbek or SE Turkmenistan	Fierce storms or major tremor
9/13–18	W. of Moscow	Powerful storms and winds
10/1–10	E. of Moscow	Fierce, destructive storms, gales
10/16–24	NE Black Sea, E. of Lugansk	Torrential rain
10/20–30	NW Ukraine to Russian border	Fierce storms, deluges

Dates	Locations	Events
10/31–11/5	NE Poland, Baltic states	Powerful storms, deluges
11/6–10	Moscow	Storms
11/13–20	E. Moscow	Powerful storms, strong winds
11/22–26	N. Sea of Azov	Storms
11/22–25	W. Russia	Storms
11/25–30	NE Caspian Sea	Storms
12/4–8	NW Siberian plain	Storms
12/17–23	NE Moscow, SW Russia, Ukraine	Storms, intense cold and snow
12/28–31	W. Russia Sumy, Kursk	Blizzards, heavy snow

Middle East

Dates	Locations	Events
1/7–12	Red Sea, Saudi Arabia	Storms, could be powerful
1/15–21	Pakistan/ Afghanistan border	Deluges, cold, snow in higher elevations
1/16–20	Saudi Arabia, Bahrain, Persian Gulf	Fierce storms, savage winds
1/20–24	S. central Turkey, Syria, Jordan, Israel, Red Sea	Storms, intense cold, rain, snow in higher elevations
1/25–29	Kuwait	Storms, strong winds
2/12–16	Egypt, N. Sudan	Storms, torrential rain, floods

Dates	Locations	Events
2/19–3/4	E. central Turkey	Storms, torrential rain, floods
3/13–22	NE Iran	Storms, torrential rain, floods, or seismic activity
3/22–31	W. Turkey	Severe storms. deluges, tremors
4/6–20	N. Turkey	Major tremors or fierce storms
4/14–22	S. Jordan, NW Saudi border	Fierce storms, powerful winds
5/6–21	Turkey	Major tremors
5/10–20	NE Oman, S. Iran	Strong tremors and/or storms
5/10–20	Arabian Sea	Tropical cyclone
6/16–24	W. Turkey	Powerful storms, torrential rain, floods
6/21–25	W. Iran, Oman	Storms, possible flash floods
6/25–30	E. Iran, Pakistan	Storms, torrential rain
7/1–16	E. Turkey	Fierce storms, deluges, major floods
8/1–12	Turkey, Syria, Saudi Arabia, Yemen	Severe storms, deluges
8/3–12	Iraq, Saudi Arabia	Destructive storms, deluges, floods, possible seismic activity
8/7–15	Iran, Gulf of Oman	Destructive storms, deluges
8/22–31	Arabian Sea, Pakistan	Destructive tropical cyclone, inundations, tidal surges
8/22–31	Kuwait, Persian Gulf	Fierce storms, powerful winds, deluges, flash floods
8/25–31	Iran (Teheran)	Major tremor
8/29–9/6	E. Egypt	Storms, could be powerful
9/12–25	N. Turkey, Armenia	Fierce storms and/or major tremors
9/17–23	Kuwait, Persian Gulf	Powerful storms
9/17–27	Iran (Tehran)	Tremors and/ or powerful storms

Dates	Locations	Events
9/21–25	NW Saudi Arabia	Powerful storms and winds
9/22–26	N. Oman/ Persian Gulf	Powerful storms and winds
9/27–10/15	E. Egypt	Fierce storms and winds, deluges
9/29–10/5	Iran	Fierce storms, powerful winds
10/1–10	SW Iran, S. Iraq	Destructive storms and winds
10/1–10	Saudi Arabia	Violent storms and winds
10/1–14	N. Turkey	Major tremor or destructive storms
10/14–22	SE Iran, Pakistan	Severe storms and winds, floods, possible seismic activity
10/20–26	N. Oman, Persian Gulf	Storms, powerful winds
10/20–30	N. Turkey	Aftershocks or renewed storms
10/27–31	SE Iran	Storms
11/12–17	Central to E. Iran	Severe weather, powerful winds
11/20–24	E. Iran	Severe weather, powerful winds
11/22–27	Central Turkey	Storms, cold weather
11/25–29	Central Iran	Mild storms
12/4–8	Central/E. Turkey	Powerful storms, low temperatures
12/11–13	Central Turkey	Storms
12/21–31	Central/E. Turkey	Severe storms, very cold, deluges, heavy snowfall in higher elevations

India, Nepal, Bangladesh, Pakistan

Dates	Locations	Events
1/6–12	NE India	Plummeting temperatures
1/17–21	Pakistan	Storms, rain

Dates	Locations	Events
1/20–23	Bangladesh	Torrential rain, cold
1/25–30	Nepal	Heavy snowfall, very cold
2/5–13	W. India	Storms and/or tremors
2/5–13	Pakistan	Cold temperatures, rain
2/21–28	Assam, Bangladesh	Storms, torrential rain
3/6–17	Assam, N. Bangladesh	Storms, torrential rain
3/18–28	Assam, N. Bangladesh	Powerful storms, rain, floods, possible tornadoes and tremor
5/4–8	SW India	Heavy rainfall, chance of tremor
5/8 –12	N.India	Fierce storms and/or powerful tremors
5/10–14	NW India, NE Pakistan	Powerful storms
5/10–30	India (W. to E.)	Horrendous, destructive storms, deluges, floods
5/18–24	Central India near Nagpur	Major earthquake or catastrophic storms
5/12–22	Bay of Bengal, E. India	Severe tropical cyclone, tidal surge, horrific winds
5/22–30	Bangladesh	Severe storms, powerful winds, deluges, floods
5/22–26	Nepal	Tremors
6/22–30	W. India, Pakistan	Torrential rain, floods
7/1–30	NW India (Rajisthan, Gujavat)	Extreme monsoon rains, Deluges, major floods
7/8–18	Bay of Bengal	Possible tropical cyclone
7/8–22	East India	Severe storms, deluges, floods

Dates	Locations	Events
7/22–30	NE India, Bangladesh	Storms, deluges, floods
8/15–18	NW India	Powerful storms, destructive winds
8/18–22	Bangladesh	Storms, floods
8/20–28	NE India	Severe storms, deluges, floods
8/21–30	NW India	Fierce storms, chance of tremor
8/22–30	S. India, Sri Lanka, Pakistan	Destructive storms, fierce winds, deluges
9/2–12	Bangladesh	Storms, rain
9/15–23	NE India, Bangladesh, Nepal	Powerful tremors and/or fierce storms, destructive winds
9/25–29	NW and SE India	Storms
9/29–10/5	W. Nepal, N. India	Torrential rains
10/2–8	S. India	Powerful storms
10/5–12	NW India	Powerful tremors
10/14–24	W. Nepal, N. India	Strong tremors and/or storms
10/20–24	Bhutan, N. Bangladesh	Powerful storms and/or seismic activity
10/24–11/3	Pakistan	Storms, deluges
10/27–11/3	W. India	Torrential rain
11/1–4	W. Nepal, NW India	Storms, torrential rain/snowfall, possible tremors
11/4–12	Assam, E. Bangladesh	Possible tremors and/or powerful storms
11/19–24	SW Pakistan	Powerful storms, torrential rain
11/21–25	SE India	Storms
11/23–26	W. Nepal, NW India	Very cold, snow in high elevations
11/24–28	N. Pakistan	Very cold, snow in high elevations

Dates	Locations	Events
12/1–4	NE India	Very cold, snow in high elevations
12/18–23	SE India, Sri Lanka	Fierce storms, high winds, deluges
12/25–31	Assam	Intense cold, deluges, snow in high elevations

China, Korea, Taiwan, Burma, Thailand, Vietnam, Laos

Dates	Locations	Events
1/16–21	Hong Kong	Severe storms, deluges, floods
1/20–23	Hong Kong	Storms, deluges, floods
1/21–25	N. Vietnam	Storms, floods
1/21–31	Central China	Severe cold, rain, snow in high elevations
1/28–31	Hong Kong	Storms
1/28–31	E. Thailand, Laos	Storms
1/28–31	N. Korea	Storms, very cold
2/5–12	Nepal	Storms, deep freeze, heavy snow
2/12–16	Burma, Laos	Storms, heavy rainfall
2/22–3/3	S. China, Hainan Dao	Powerful storms, heavy rainfall
2/28–3/3	Burma, Laos	Storms, heavy rainfall, floods
3/8–16	E. China near Shang-hai	Powerful storms and/or powerful tremors
3/13–20	NE China	Tremors, could be powerful
3/17–25	E. China near Shang-hai	Powerful storms and/or powerful tremors
3/20–26	E. Thailand	Powerful storms, fierce winds

Dates	Locations	Events
3/20–28	NE of Hainan Dao	Powerful storms or tremors
3/30–4/4	E. Thailand	Powerful storms, torrential rain
4/2–14	Hainan Dao	Major quakes or powerful storms
4/5–14	NE China	Powerful quake
4/11–22	S. China, Burma	Fierce storms, powerful winds, floods
4/11–22	S. China Sea	Possible typhoon
4/14–20	SE China	Powerful storms, major floods, and/or powerful quakes
4/18–26	Hong Kong	Powerful storms, major floods, or major quakes
4/26–5/7	W. China	Storms, torrential rain, floods
5/4–12	NE China	Storms, torrential rain
5/6–14	SE China, Taiwan	Major quakes
5/20–24	Mongolia, central China, SE Asia	Fierce, destructive storms, gale-force winds
6/2–14	Central/E. China, Burma, Thailand	Storms, could be powerful
6/16–7/8	Central/E. China	Powerful storms and floods
8/15–20	S. Thailand, Burma	Powerful storms, fierce winds
8/25–31	Andaman Sea	Fierce typhoon
8/25–31	Japan	Fierce typhoon
8/26–31	N. Korea	Powerful storms
8/29–31	S. China, N. Vietnam	Fierce, destructive storms
8/28–31	E. Mongolia, NE China	Fierce storms, possible tremors
9/12–23	E. Mongolia, China	Storms
9/17–23	S. China, N. Vietnam	Fierce storms

Dates	Locations	Events
9/17–23	NW China	Powerful storms and/or tremors
9/27–29	E. China	Storms
10/1–10	E. Taiwan	Severe typhoon
10/1–10	E. Mongolia, NE China	Fierce storms and/or major quakes
10/15–22	NE China, N. Korea	Fierce storms or powerful quakes
10/15–22	S. China	Storms
10/20–24	Bhutan, S. China	Storms
10/23–28	Central China	Severe storms
11/1–3	Bhutan, S. China	Storms
11/18–25	NW China, S. Mongolia	Storms, very cold, torrential rain or snow
12/5–8	Hainan Dao, S. China	Storms or strong tremors
12/18–25	Central Mongolia	Severe winter storms, deep freeze, snow and possible blizzards
12/27–31	S. Burma, Thailand	Fierce storms and floods

Indonesia and Papua New Guinea

Dates	Locations	Events
1/6–15	Irian Jaya	Storms, torrential rain
1/15–21	Borneo, Irian Jaya	Severe storms, deluges, floods
1/17–25	Papua New Guinea	Severe storms, deluges, floods
1/21–26	Borneo	Storms, deluges, floods
1/30–2/2	S. Sumatra, N. Java	Torrential rain, floods
2/5–8	Borneo	Torrential rain, floods

Dates	Locations	Events
2/12–15	Celebes	Storms
2/16–25	Papua New Guinea	Offshore quakes
2/27–3/4	Araufura Sea	Tropical cyclone
2/28–3/4	Celebes	Storms, torrential rain
3/13–16	Sumatra	Storms
3/18–30	Sumatra	Powerful storms and possible seismic activity–could be powerful
3/22–4/4	Papua New Guinea, New Britain	Powerful storms, floods, possible tremors
3/28–4/2	Sumatra	Storms
3/28–4/2	Gulf of Thailand	Storms
3/29–4/2	South of Guam	Offshore quakes
3/30–4/12	Java, Borneo	Powerful tremors, volcanic activity, or major inundations
4/12–22	Borneo	Severe storms, powerful winds, torrential rain, major floods
4/19–23	Sumatra	Severe storms, powerful winds, possible tremors
5/8–22	Sumatra, Borneo, Rabual	Powerful earthquakes, severe weather, volcanic activity Anuk Krakatoa or Rabual's volcano
5/17–24	Arafura Sea, Gulf of Carpentaria, Lesser Sunda Is.	Catastrophic storms, possible cyclone and storm surge
6/16–25	Sumatra, Irian Jaya, Papua New Guinea	Powerful storms, torrential rain, floods
6/16–28	Borneo, Celebes	Powerful storms, torrential rain, floods
7/1–8	N. of Celebes, S. of Mindanao	Horrendous storms, typhoon, or major seaquake and tsunami

Dates	Locations	Events
7/8–16	N. Moluccan Is. Halmehara	Severe storms, powerful winds, floods
7/16–30	New Guinea, New Britain	Powerful storms, torrential rain, floods
7/27–8/12	S. Sumatra, W. Java	Horrendous storms, major floods, volcanic activity Anuk Krakatoa
8/6–16	Papua New Guinea, New Britain, Bougainville	Horrendous storms, floods, or powerful quakes and/or eruption of Rabual's volcano
8/15–20	N. Sumatra	Severe storms, powerful winds
8/15–9/2	Andaman Sea, E. Philippines	Powerful typhoons
8/22–31	Papua New Guinea	Destructive storms, floods, or powerful tremor
8/24–31	N. Sumatra	Severe storms, typhoon
9/11–22	N. Sumatra	Powerful storms, inundations, and/or seismic activity
9/13–22	S. Sumatra, W. Java	Powerful storms, inundations
9/17–23	N. Philippines	Fierce typhoon
9/25–10/6	Borneo	Major quake and/or powerful storms
9/26–10/10	W. Irian Jaya	Destructive storms, fierce winds
9/27–10/5	Celebes, Timor	Powerful, destructive storms
10/1–10	NW Pacific	Fierce typhoons
10/2–8	Java	Fierce, destructive, storms, gale force winds, possible tremors
10/13–22	N. Borneo	Powerful storms
10/23–31	Borneo	Powerful storms
10/27–11/3	New Britain	Torrential rain, floods

Dates	Locations	Events
10/27–11/6	Irian Jaya, Papua New Guinea	Torrential rain, floods
11/4–6	Borneo	Storms
11/6–10	N. Sumatra	Offshore storms
11/13–17	N. Sumatra, New Britain	Storms
11/20–24	Sumatra, New Britain	Powerful storms, floods
11/22–25	Moluccas	Storms
11/25–29	Irian Jaya, Papua New Guinea	Storms
12/4–8	Celebes, Sumatra, Java	Storms, torrential rain
12/10–13	Celebes	Torrential rain
12/18–25	Irian Jaya, Papua New Guinea	Fierce storms, destructive winds
12/28–31	Lesser Sunda Is., Lombok, Bali	Fierce storms, torrential rain, floods

The Philippines

Dates	Locations	Events
1/16–21	NE Luzon	Severe, destructive storms, floods
2/12–16	Philippines	Storms
2/27–3/4	Philippines	Storms, torrential rain, floods
3/10–20	SE Luzon and nearby Pacific	Major tremors and/or severe weather with major floods
3/30–4/4	Palawan	Storms, torrential rain, floods
4/10–22	Palawan	Major quake and/or major floods

Dates	Locations	Events
4/18–22	N. Luzon	Fierce storms or powerful tremors
4/28–5/3	N. Luzon	Fierce storms or powerful tremors
5/3–10	Sulu Archipelago	Powerful storms
5/7–12	N. Luzon	Major floods or major offshore quake
5/17–24	Sulu Archipelago	Fierce, destructive storms, gales
6/18–30	Luzon	Storms, torrential rain
7/1–16	E. Philippines, Samar, Mindanao	Catastrophic storms, deluge, possible offshore quake
8/24–30	Luzon	Powerful, destructive storms
9/17–23	N. Luzon	Powerful typhoon
10/2–11	Mindanao	Destructive storms, fierce winds, and/or major tremors
12/27–31	Sulu Archipelago	Powerful storms, floods

Japan

Dates	Locations	Events
1/6–18	Honshu	Heavy snows
1/17–21	Hokkaido	Fierce storms, heavy snow, blizzards
1/20–25	Kyushu	Storms, could be powerful
2/18–23	Honshu	Storms
2/28–3/3	Kyushu, S. Honshu	Storms, deluges
3/13–22	Honshu	Deluges, floods
3/26–4/4	Honshu	Storms, deluges, floods
4/4–14	Kuril Is.	Powerful seismic activity
5/3–12	Honshu, Hokkaido	Powerful storms, fierce winds

Dates	Locations	Events
5/16–24	N. Honshu, Hokkaido	Fierce storms, gale-force winds from nearby typhoon and/or tremors
5/27–6/4	Kyushu	Powerful storms, floods
6/14–20	Honshu	Storms, torrential rain, floods
6/18–24	Hokkaido	Storms, torrential rain, floods
7/1–10	N. Honshu, Hokkaido	Powerful storms, inundations
7/16–25	Honshu	Torrential rain, possible storms
7/31–8/15	Honshu	Powerful storms, inundations or powerful seismic activity
8/22–31	N.Honshu, Hokkaido, Kyushu	Horrific storms, gale-force winds, and/or powerful earthquake
8/22–31	Kyushu, Hokkaido	Fierce typhoons
9/14–25	Kyushu, Hokkaido	Powerful storms, fierce winds, and/or powerful seismic activity
9/15–23	Kyushu, Hokkaido	Fierce typhoons
10/1–10	S. of Japan	Fierce typhoon
10/1–14	Tokyo, SE Honshu	Powerful typhoon or powerful earthquake or volcanic activity
10/14–24	Honshu	Fierce destructive typhoon
10/27–11/2	Honshu	Storms, torrential rain
11/3–14	Honshu	Torrential rain
11/19–25	N. Honshu	Fierce storms, powerful winds, possible seismic activity
11/28–12/1	N. Honshu	Storms
12/18–23	Honshu	Powerful storms, cold, heavy snow
12/25–31	Kyushu	Powerful storms, torrential rain

New Zealand

Dates	Locations	Events
1/19–26	SE South Island	Deluges, floods
1/28–31	North Island	Storms, torrential rain
2/5–15	South Island	Strong tremors
2/19–28	North Island	Powerful storms, strong tremors
3/10–30	North Island	Powerful tremors, possible volcanic activity, storms
3/30–4/2	South Island	Powerful storms. deluges
4/8–14	N. South Island	Powerful storms, seismic activity
4/11–22	NE North Island	Powerful tremors, possible volcanic activity, storms
5/4–12	North Island	Powerful tremors, volcanic activity, or storms, deluges
5/10–18	South Island	Powerful storms, deluges, floods, or powerful tremors
6/2–7	North Island	Storms, deluges
6/18–25	Both islands	Storms, torrential rain, floods
6/28–7/8	South Island	Major floods , powerful storms
7/23–30	South Island	More floods
7/31–8/12	North Island	Horrific storms, gale-force winds, major floods, and/or powerful tremors and volcanic activity
8/14–20	Both islands	Fierce storms and winds, possible seismic activity
8/22–9/3	Both islands	Horrific storms, powerful winds, seismic or volcanic activity
9/17–23	Both islands	Powerful storms and winds
10/2–10	W. of South Island	Fierce destructive storms
10/7–14	South Island	Storms

Dates	Locations	Events
10/18–24	South Island	Powerful storms, winds, deluges
10/14–20	North Island	Powerful storms, seismic activity
11/4–16	Both islands	Major tremor/volcanic activity
11/19–24	NE North Island	Major tremors, powerful storms
11/22–27	South Island	Storms
12/2–6	South Island	Deluges, storms, chance of tremor
12/17–23	S. South Island	Fierce storms, winds, deluges
12/23–31	North Island	Powerful storms, deluges

Australia

Dates	Locations	Events
1/17–21	Western Gulf, Arnhem Land	Tropical cyclone
1/17–21	SE Queensland, Brisbane	Fierce storms, deluges
1/17–25	Coral Sea	Tropical cyclone
1/20–25	SW Western Australia	Deluges
2/5–8	E. New Caledonia	Tropical cyclone
2/18–20	S. Victoria, Melbourne	Storms
2/28–3/4	Coral Sea	Tropical depression, cyclone
2/28–3/4	SE Queensland, Brisbane N. NSW	Storms,deluges
3/10–20	SE Queensland, Brisbane, N. NSW	Deluges, major floods

Dates	Locations	Events
3/20–28	Coral Sea	Powerful tropical cyclone
3/20–28	Melville Island, Nortern Territory	Powerful tropical cyclone
3/26–31	Gulf region	Storms, deluges
3/28–31	SE Queensland, Brisbane	Storms, deluges
4/5–10	NE Queensland	Storms, deluges
4/11–14	SE Queensland	Storms
4/13–22	Off Exmouth	Tropical cyclone
5/3–8	Great Sandy Desert, Canning Basin	Storms
5/4–13	Central NSW, Queensland	Fierce storms, powerful winds
5/5–10	Barkly Tableland, N. Queensland	Storms
5/10–14	Coral Sea	Tropical cyclone
5/12–16	SE Queensland, Brisbane	Powerful storms
5/12–22	S. Pacific, Fiji	Destructive cyclone, gale-force winds, storm surges
5/14–22	NSW, Sydney	Fierce, destructive storms, gales
5/18–22	Offshore NSW	Destructive storm, gales
5/19–24	SE Queensland, Brisbane	Fierce, destructive storms, gales
6/26–7/16	NSW, Sydney	Powerful storms, deluges, major floods
7/8–16	S. Pacific, New Caledonia	Severe storms, deluges, floods
7/16–31	Victoria to SE Queensland	Torrential rain

Dates	Locations	Events
7/31–8/4	S. Victoria, Melbourne	Storms, torrential rain
8/5–16	SE South Australia, SW Victoria, N. Tasmania	Fierce storms, winds, floods
8/18–26	SE South Australia, SW Victoria	Powerful storms
8/22–26	E. NSW	Powerful storms, deluges
8/26–31	NE NSW, SE Queensland	Fierce storms, gales
9/10–17	SW Victoria, Melbourne	Powerful storms, deluges
9/13–20	Central W. NSW, NW Victoria	Powerful storms.deluges
9/15–23	Great Sandy Desert,	Fierce storms, powerful winds
9/21–26	Arnhem Land, Hamersley Range	Storms
9/22–29	SW Pacific, New Caledonia	Fierce storms, powerful winds
10/2–10	Arnhem Land, Barlee Ranges	Fierce, destructive storms, gales, and floods
10/2–10	Central W. NSW, NW Victoria	Fierce, destructive storms, gales, floods
10/6–11	S. Victoria	Storms
10/13–17	SW Pacific, New Caledonia	Fierce storms, powerful winds
10/13–17	Coral Sea off NSW	Fierce storms
10/13–17	Dirk Hartog Is.	Fierce storms
10/14–22	S. Victoria	Torrential rains, possible storms
10/23–26	Cape York	Torrential rain

Dates	Locations	Events
10/30–11/15	Arnhem Land, Barkly Tableland	Deluges, floods
10/30–11/15	Central Queensland	Deluges, floods
11/4–15	E. NSW	Torrential rain periods
11/13–18	Cape York	Powerful storms
11/19–24	N. Queensland	Powerful storms, torrential rain
11/25–28	W. Australia, Kimberley Plateau	Storms
12/1–4	W. Australia, Kimberley Plateau	Storms
12/18–24	Gulf of Carpentaria	Tropical depression, cyclone
12/18–24	E. of Vanuatu	Tropical cyclone
12/18–25	S. Victoria	Storms, torrential rain
12/21–25	SE Queensland	Powerful storms, deluges
12/28–31	Far NE Queensland	Powerful storms, deluges

Hawaiian Islands

Dates	Locations	Events
1/16–23	Kauai, Oahu	Deluges, floods
1/17–23	Big Island	Storms, deluges
1/28–30	Kauai	Torrential rain
2/4–10	Kauai, Oahu	Tremors and/ or powerful storms
2/13–16	Big Island	Storms
2/19–25	Big Island	Seismic activity and storms
2/22–3/2	Kauai, Oahu	Storms and possible tremors
3/22–25	Big Island	Storms and/or seismic activity
3/28–4/12	Kauai to Big Island	Powerful tremors, powerful storms

Dates	Locations	Events
4/24–28	Kauai, Oahu	Storms
4/30–5/12	Big Island	Powerful storms, torrential rain, floods
5/14–24	Big Island	Powerful storms, fierce winds, possible tremors/volcanic activity
5/26–6/1	Kauai, Oahu	Storms, could be powerful
6/6–16	Kauai to Big Island	Powerful storms, strong winds, floods
6/18–30	Kauai to Maui	Powerful storms, torrential rain, floods
7/1–8	W. of Hawaii	Fierce storms, possible hurricane
8/5–12	Central Pacific	Powerful hurricane
8/15–17	Kauai, Oahu	Storms, powerful winds
8/22–30	Kauai to Big Island	Powerful storms, fierce winds, possible seismic activity
8/22–30	Central Pacific	Powerful hurricane
9/13–21	Big Island	Powerful storms, fierce winds
10/20–23	Kauai, Oahu	Storms
10/28–31	Kauai, Oahu	Storms
11/20–25	Big Island	Storms
12/10–12	Kauai	Torrential rain
12/21–25	Kauai, Oahu	Storms

Alaska

Dates	Locations	Events
1/25–28	Anchorage	Heavy rainfall or snow
2/12–20	Anchorage	Storms, heavy rain or snow
3/13–17	Gulf of Alaska	Torrential rain, possible storms
3/20–28	Central Alaska, Anchorage	Powerful storms, strong winds

Dates	Locations	Events
4/4–12	Kodiak Is.	Major tremors
4/10–20	Anchorage	Major tremors
5/18–23	Kodiak Is., Gulf of Alaska	Fierce, destructive storms
6/16–28	NE Pacific Aleutian Islands	Powerful storms
7/16–24	NE Pacific Aleutian Islands	Powerful storms, torrential rain
8/7–10	S. of Kodiak Is.	Storm activity
8/15–20	Kodiak Is.	Powerful tremors and/or storms
8/26–31	Kodiak Is.	Powerful storms
9/17–22	Gulf of Alaska	Storm activity
10/13–18	S. Aleutian Islands	Tremors
11/1–8	N. Alaskan Peninsula, Kodiak Is.	Storms, deluges
11/22–25	Gulf of Alaska	Storms, deluges
12/5–8	S. Aleutian Islands	Seismic activity
12/27–31	S.Aleutian Islands	Powerful storms and/or powerful seismic activity

U.S.A.: West Coast

Dates	Locations	Events
1/16–20	N. California	Powerful storms, deluges
1/18–23	S. California	Powerful storms, deluges
1/29–2/1	S. California	Storms
2/10–16	S. California	Storms
2/18–26	N. California	Tremors, could be powerful

Dates	Locations	Events
3/13–18	N. California	Deluges, floods
3/19–26	W. Washington, W. Oregon	Storms, torrential rain, floods
3/22–30	N. California	Storms, floods
3/26–4/2	S. California	Storms, floods
4/26–5/5	California	Torrential rain
5/5–8	California	Storms
5/7–22	N. California, San Francisco	Major earthquake, esp. 5/8–12 and 5/19–22
5/18–26	S. California	Torrential rain
6/14–18	S. California	Torrential rain
7/8–18	West Coast, esp. California	Powerful storms, major floods
7/22–30	N. California	Deluges and floods
7/25–31	S. California	Deluges and floods
8/1–3	S. California	Storms
8/4–12	S. California desert, Baja Peninsula	Fierce storms, deluges, floods, possibly hurricane related
8/22–27	N. California	Storms
8/27–31	S. California	Storms
9/14–19	N.California	Powerful storms
10/2–10	Oregon, possibly N. California	Severe storms, extreme winds, or powerful earthquakes
10/2–10	Baja peninsula	Hurricane targets or forms
10/9–14	N. California	Fierce storms, powerful winds
10/17–24	N. California	Fierce storms, powerful winds
10/27–30	S. California	Torrential rain
10/29–11/4	N. California	Storms

Dates	Locations	Events
11/12–17	Oregon, N. California	Tremors
11/18–23	S. California	Powerful storms
12/18–23	California	Powerful storms, deluges
12/28–31	N. California	Fierce storms, deluges

Continental U.S.A.

Dates	Locations	Events
1/5–10	S. Ontario, Lake Superior, Michigan, Wisconsin	Cold, heavy snowfall
1/10–12	Washington state	Heavy rainfall or snowfall
1/14–18	Arkansas, Missouri	Storms, torrential rain
1/15–20	Washington state, Oregon	Powerful storms, deluges, snow
1/14–19	W. Texas	Powerful storms, deluges
1/20–25	Upstate New York, Vermont, Wisconsin, Mississippi Valley	Deluges, snow
1/28–31	Mississippi, SW Tennessee	Deluges
1/28–31	Washington state, Oregon, Wyoming N. California, Colorado	Deluges, snow
1/28–31	Virginia, Delaware, New Jersey	Storms, deluges, snow
2/5–8	Colorado	Tremors

Dates	Locations	Events
2/4 –8	New Madrid fault or nearby states	Strong tremors
2/5–10	Louisiana, Mississippi, Tennessee	Torrential rain
2/12–16	New England	Torrential rain, possible storms
2/13–16	North Carolina, Virginia	Storms, deluges
2/20–25	Maine, New Brunswick	Heavy rainfall
2/21–25	New Mexico	Deluges
2/27–3/2	Maine	Storms, deluges
2/27–3/4	Alabama, Tennessee, Kentucky	Storms, deluges, floods
3/6–10	Mississippi, Alabama, Tennessee	Deluges and floods
3/10–20	Louisiana	Storms, torrential rain, floods
3/10–13	N. and W. Texas	Storms
3/13–20	Washington state, Oregon	Storms, deluges, floods
3/18–24	W. Texas	Fierce storms, tornadoes
3/22–28	Mississippi, Alabama, Tennessee	Deluges, floods
3/24–27	Oregon	Powerful storms, deluges
3/28–4/1	Alabama, Florida	Storms, torrential rain
3/28–4/1	Utah, Arizona	Deluges, possible storms
3/30–4/2	Washington state, Oregon	Powerful storms, deluges, floods

Dates	Locations	Events
4/5–10	Wyoming, Colorado, possibly New Mexico	Tremors and/or torrential rain
4/6–14	New York, New England	Deluges, major floods, or powerful tremors
4/10–14	Texas	Powerful storms, tornadoes
4/11–14	Central Montana, NW Wyoming	Powerful tremors
4/16–22	Nebraska, Kansas	Powerful storms, tornadoes
4/18–22	Maine	Fierce storms, powerful winds
4/18–22	Wyoming, Dakotas, Colorado, possibly New Mexico	Fierce storms, tornadoes
4/18–22	E. coast, Carolinas and points north	Powerful storms, tornadoes
4/19–22	Tennessee, Alabama, possibly Florida	Powerful storms, tornadoes
4/22–26	E. Texas, Louisiana	Powerful storms, tornadoes
4/27–30	Louisiana	Torrential rain
4/28–5/4	Missouri, Iowa, Illinois, Kentucky	Storms, deluges, tornadoes
4/28–5/4	Carolinas	Storms and torrential rain
5/1–6	Texas, Wyoming, Colorado	Torrential rain
5/3–10	NE states	Powerful storms
5/3–10	N. Montana	Powerful storms, deluges
5/7–14	Texas	Powerful storms, tornadoes
5/12–24	Carolinas to Washington, D.C.	Fierce destructive storms, gale-force winds, tornadoes
5/17–24	Texas	Fierce destructive storms, gale-force winds, tornadoes

Dates	Locations	Events
5/18–24	Dakotas	Fierce storms, destructive winds, possible tornadoes
5/20–26	NE U.S.A.	Fierce destructive storms
5/24–31	E. Texas, Louisiana, Mississippi, Florida, Alabama	Powerful storms, destructive winds, possible tornado outbreaks
5/26–31	Dakotas, Manitoba	Powerful storms, destructive winds
5/26–6/4	Indiana, Kentucky, Tennessee	Powerful storms, destructive winds
5/27–6/4	Minnesota	Powerful storms, destructive winds
5/30–6/5	Alabama, Florida	Powerful storms, destructive winds possible tornadoes
6/2–4	NW Wyoming, S.Montana	Storms, powerful winds
6/10–20	Indiana, Kentucky, Tennessee, Kansas, Minnesota, Dakotas	Storms, deluges, floods
6/18–25	Alabama, Missouri, Tennessee, Florida	Storms, deluges, floods (Mississippi river could flood)
6/21–27	Texas, Oklahoma, Arkansas, Missouri	Storms, deluges, floods
6/26–7/14	Mississippi and Ohio rivers	Major floods
6/28–7/10	East Coast	Storms, deluges, floods
7/1–10	Mexico, Texas, Arkansas	Powerful storms, possibly hurricane relared
7/25–8/2	Texas, Louisiana, Montana, Nebraska, Dakotas, Kansas	Storms, torrential rain, floods
7/29–8/2	Louisiana, Mississippi	Storms, torrential rain, floods

Dates	Locations	Events
8/4–12	Montana, Wyoming, N. Dakota, Florida, Texas, Minnesota	Fierce destructive storms, deluges, floods, possible tornadoes
8/12–18	Wyoming, Colorado, N. Dakota, Texas, Minnesota, Illinois, Missouri	Fierce, destructive storms, deluges, floods, possible tornadoes
8/16–31	Gulf or Caribbean	Powerful hurricanes
8/21–24	Montana, Wyoming, Colorado, Dakotas, NW Nebraska, Texas, Missouri, Minnesota, Florida, Illinois	Storms, torrential rain
8/25–31	Florida, Carolinas	Offshore hurricane
8/25–31	Dakotas, Minnesota, Michigan, Iowa, Missouri, Kansas, Nebraska	Fierce storms, destructive winds possible tornadoes
8/25–31	Texas, New Mexico	Strong storms
9/15–24	N. Carolina and points north	Offshore hurricane
9/13–20	Texas	Powerful storms, winds, deluges
9/15–20	N. Carolina to New York	Powerful storms, winds, deluges
9/15–20	N. and S. Dakota, New Mexico, Idaho, W. Montana	Powerful storms, winds, deluges
9/15–23	S. Florida	Destructive storms, powerful winds, deluges, floods
9/25–27	New York, New Jersey	Storms
9/30–10/3	New York	Deluges, floods

Dates	Locations	Events
10/2–10	Texas, Louisiana, Oklahoma, and/or W. Kansas	Fierce destructive storms, deluges, tornadoes
10/2–10	Gulf of Mexico	Hurricane
10/5–10	S. Arizona, upper Mexico	Destructive storms, possible tremors
10/7–9	New England	Torrential rain
10/23–11/3	S. Carolina and points north	Powerful storms
10/24–30	Nevada, Utah	Storms, plummeting temperatures
10/27–31	Mississippi, Florida, Alabama	Powerful storms
10/28–11/10	Missouri, Arkansas	Powerful storms, deluges
10/30–11/5	Texas	Powerful storms, deluges
11/3–10	N. Dakota, Minnesota	Powerful storms
11/4–14	Mississippi, Alabama, NW Florida	Fierce storms, deluges
11/6–12	Nevada	Fierce storms or powerful tremors
11/6–15	S.tip of Florida	Fierce storms, deluges, floods
11/11–18	Washington state, Oregon	Powerful storms
11/18–24	Arizona	Powerful storms, cold, deluges
11/19–24	N. Florida, SW Georgia	Fierce storms, deluges
11/19–24	Washington state, Oregon	Powerful storms, deluges
12/1–4	Washington state, Oregon	Deluges, cold

Dates	Locations	Events
12/1–5	Wyoming, Colorado	Plummeting temperatures, deluges, snow
12/4–8	Maine	Plummeting temperatures, snow
12/4–8	Idaho, Montana	Plummeting temperatures, snow
12/5–8	Louisiana	Torrential rain
12/18–20	New England	Winter storms
12/20–25	Illinois, Missouri, Arkansas, W. Tennessee	Storms
12/21–25	Ontario, Michigan, Lake Superior, Wisconsin	Fierce winter storms, blizzards
12/25–28	S. Florida	Could bring a "snap freeze"
12/27–31	Mississippi, Louisiana	Fierce storms, deluges, floods
12/27–31	NW Montana, N. Idaho	Powerful storms, very heavy snow, possible blizzards

Mexico

Dates	Locations	Events
1/9–11	Campeche	Cold and rainy
1/14–18	Yucatan	Cold and rainy
1/20–28	Tabasco, Yucatan, Guatemala	Storms, deluges
1/28–2/7	Offshore SW Mexico	Fierce storm systems
2/5–8	W. Gulf of Mexico	Storms
2/19–23	Tabasco, Campeche, Yucatan	Storms

Dates	Locations	Events
2/22–28	Durango, Zacatecas, Guadalajara	Strong tremors
2/27–3/4	Yucatan Peninsula	Storms, deluges, floods
3/9–18	Chiapas, Campeche	Powerful tremors and/or storms, major floods
3/10–16	Yucatan Peninsula	Deluges, major floods
3/10–18	Mazatlan	Storms, deluges
3/18–28	Yucatan Peninsula,	Powerful storms, deluges, major floods
3/18–28	Michoachan	Powerful tremors, powerful storms
3/24–28	NW plateau	Tremors
4/5–12	Mazatlan	Powerful tremors
4/11–18	Vera Cruz, Tabasco	Storms, deluges, and/or tremors
4/11–18	W. Gulf of Mexico	Storms
4/16–22	Michoacan, W. Mexican plateau	Powerful tremors, powerful storms
4/26–30	E. Oaxaca, Minatitlan	Fierce storms, strong winds, deluge
5/5–8	Baja Peninsula	Powerful storms
5/8–14	Mexico City, San Luis Potosi	Major tremors
5/17–24	Mexico City, San Luis Potosi, Tampico	Powerful earthquakes, volcanic activity or fierce storms
5/19–26	Baja Peninsula	Fierce storms, powerful winds, good chance of powerful tremors
6/2–6	Tamaulipas	Storms
6/21–31	S. and SW Mexico	Storms, deluges, floods
6/25–7/5	Tabasco, Yucatan Peninsula	Storms, deluges, floods

Dates	Locations	Events
6/25–7/3	Maztlan, Tuxpan	Hurricane, severe floods
7/1–10	Gulf of Mexico	Hurricane, severe floods
7/1–20	S. Baja, Durango	Powerful storms, massive deluges, severe floods
7/24–8/6	Yucatan Peninsula	Storms, deluges, major floods
7/31–8/12	Baja Peninsula, Los Mochis region, Durango, Navarit	Horrendous storms, fierce winds, widespread and severe floods
8/5–13	Pacific coast near Navarit or Durango	Fierce hurricane, severe floods
8/15–24	Pacific coast near Mazatlan	Powerful hurricane, severe floods
8/15–24	Gulf of Mexico, Yucatan Peninsula	Powerful hurricane
8/15–24	El Dorado, Culiacan Rosales	Powerful storms, severe winds, deluges,floods
8/21–28	Mexico City	Powerful storms, deluges and floods, possible seismic activity
8/21–26	E. Oaxaca	Storms, deluges, floods
8/22–26	Baja Peninsula, Guerrero	Powerful storms, deluges, floods
8/24–30	Pacific coast near Colima, Michoacan	Powerful hurricane, floods
8/26–9/3	Gulf of Mexico, Yucatan Peninsula	Fierce hurricane
8/26–9/1	Pacific coast near Mazatlan	Fierce hurricane
9/15–23	NW Pacific	Hurricane
9/16–24	Gulf of Mexico, Campeche, Yucatan	Tropical storm
9/25–28	Mazatlan	Storms, deluges

Dates	Locations	Events
9/29–10/3	Navarit	Deluges, floods
10/2–10	W. Oaxaca	Powerful earthquake
10/2–10	NE Pacific	Hurricane
10/5–18	Baja Peninsula	Hurricane or powerful tremors
10/15–23	N. Tiaxcala	Powerful storms and/or tremors
10/20–25	Mazatlan	Powerful storms, deluges, floods, possible seismic activity
10/20–26	Yucatan Peninsula	Powerful storms, deluges, floods
10/28–31	N. Baja Peninsula	Storms
11/1–4	Hidalgo	Powerful storms, deluges, floods
11/7–11	Yucatan Peninsula	Powerful storms, deluges, floods, possible seismic activity
11/22–26	Yucatan Peninsula	Storms, deluges
12/1–4	Chiapas	Storms, deluges, cold
12/18–23	Baja Peninsula	Powerful storms, deluges, floods
12/26–30	Baja Peninsula	Powerful tremors
12/27–31	Yucatan Peninsula	Powerful storms, deluges, floods

Canada

Dates	Locations	Events
1/6–10	W. Ontario	Cold, heavy snowfall
1/20–22	British Columbia	Storms, deluges, or snow
1/20–24	Lake Superior	Deep freeze, heavy snowfall
1/21–24	SE Quebec, Gulf of St. Lawrence, New Brunswick, Nova Scotia	Deep freeze, heavy snowfall

Dates	Locations	Events
1/29–2/1	SW Quebec	Storms, deluges, or snow
2/5–7	James Bay, W. Quebec	Severe cold, storms, heavy snow
2/12–14	NW Brit. Columbia	Possible storms
2/16–19	NW Brit. Columbia	Storm activity, snow
2/23–27	SE Saskatchewan New Brunswick	Storms, snow
2/27–28	E. Ontario	Storms, snow
3/10–20	NE Ontario	Deluges, floods
3/13–18	SE Quebec	Storm activity
3/14–20	E. Brit. Columbia	Powerful storms, deluges, floods
3/22–28	S. Saskatchewan	Powerful storms and winds
3/22–30	Alex. Archipelago	Powerful storms
3/23–28	NE Saskatchewan, Manitoba	Powerful storms, deluges, floods
3/28–4/1	NW Alberta	Storms and torrential rain
3/30–4/6	Nova Scotia	Storms and torrential rain
4/5–16	SE Quebec	Powerful tremors or storms, floods
4/14–21	SE Saskatchewan	Powerful storms, deluges
4/18–21	SE Quebec	Powerful tremors or storms, floods
5/3–10	SE Alberta, Saskatchewan	Deluges, storms, floods
5/4–8	Nova Scotia	Deluges
5/18–25	Manitoba	Fierce storms, winds, torrential rain, possible floods
5/26–31	N. Alberta	Powerful storms, fierce winds, torrential rain, possible floods
6/2–6	NW Brit. Columbia	Powerful storms, fierce winds

Dates	Locations	Events
7/31–8/11	Saskatchewan, W. Manitoba	Fierce storms, winds, deluges, floods
8/7–15	S. Ontario	Powerful storms, fierce winds, deluges, floods
8/11–16	Manitoba	Powerful storms, deluges, floods
8/13–20	SW Ontario, Toronto	Powerful storms
8/26–31	SE Ontario, SW Quebec	Fierce storms, very high winds
8/26–31	Vancouver Is. to SE Alberta	Severe, powerful storms
9/13–20	SE Brit. Columbia	Fierce storms, high winds, deluges
10/2–10	Newfoundland Is.	Strong storms, fierce winds
10/5–12	NE Brit. Columbia, Yukon, NW Alberta	Powerful storms
10/5–15	Saskatchewan	Fierce storms, powerful winds
10/13–18	Ontario	Powerful storms, intense cold
10/17–23	NE Brit. Columbia, Yukon, NW Alberta	Fierce storms, snap freeze
10/27–31	Ontario	Powerful storms
11/3–6	Vancouver Is., SW Brit. Columbia	Deluges, dropping temperatures
11/13–16	SE Brit. Columbia, Alberta	Storms, cold
11/18–24	SW Manitoba	Fierce storms, powerful winds, snow
11/19–25	Brit. Columbia	Powerful storms
11/22–27	Newfoundland Is.	Storms
11/25–29	Vancouver Is.	Storms, torrential rain
12/3–6	S. Saskatchewan, Vancouver Is.	Deluges and/or heavy snowfall

Dates	Locations	Events
12/18–25	Alex. Archipelago, W. Brit. Columbia	Powerful storms, snow

Central and South America, Greater and Lesser Antilles

Dates	Locations	Events
1/15–21	Venezuela	Fierce storms, winds, deluges
1/16–21	NE Brazil	Fierce storms, winds, deluges
1/17–21	Lesser Antilles, Leeward Is.	Fierce storms, winds, deluges
1/21–27	W. Brazil, Rondonia	Deluges
1/21–29	Yucatan, Guatemala	Deluges
1/28–30	Nicaragua, Brazil, French Guyana	Deluges
1/30–2/1	W. Columbia	Storms, deluges, floods
2/4–6	E. Brazil, W. Gerais	Storms
2/12–16	SE Pacific, N. Peru	Ocean storms
2/12–16	Lesser Antilles, Windward Islands, N. Brazil	Storms, deluges
2/21–28	NW Bolivia	Powerful tremors and/or storms
2/23–3/1	Yucatan, Honduras, El Salvador, Belize	Powerful storms, deluges, floods
2/27–3/2	E. Brazil, Goias	Storms, torrential rain, floods
2/27–3/4	NE Chile, S. Bolivia	Powerful storms, deluges, floods
3/11–16	E. Brazil, N. Bahia	Storms
3/12–20	Honduras, Guatemala, El Salvador	Torrential rain, floods

Dates	Locations	Events
3/13–20	Panama	Tremors and/or storms
3/13–24	Uruguay, SW Brazil	Powerful earthquakes and/or fierce storms, deluges, floods
3/15–18	S. Chile	Storms
3/18–26	N. Columbia	Fierce storms, powerful winds, and/or seismic activity
3/18–28	SW Brazil	Destructive storms, major floods, or powerful tremors
3/20–28	S. Bolivia to Argentina border	Fierce, destructive storms, major floods, or powerful tremors
3/26–4/1	Honduras, Costa Rica	Storms, deluges
3/30–4/8	Leeward Islands	Powerful storms, deluges, possible volcanic activity
3/30–4/2	S. Chile, N. Bolivia, Rondonia	Powerful storms, deluges
3/30–4/8	E. Venezuela, N. Brazil	Powerful storms, deluges, floods
4/5–12	S. Guyana, N. Brazil, S. Columbia, N. Peru	Storms, deluges and/or tremors
4/10–22	Central Chile, Argentina border	Powerful tremors and/or fierce storms
4/11–14	Panama	Tremors or powerful storms
4/12–21	S. Bolivia, Argentina border	Destructive storms, major floods
4/13–21	Lesser Antilles	Powerful storms, deluges, volcanic activity
4/16–22	Panama	Powerful tremors or fierce storms and floods
4/18–22	E. Puerto Rico, S. Cuba	Fierce offshore storms or quakes

Dates	Locations	Events
4/22–26	Argentina	Powerful storms
5/2–20	Lesser Antilles	Possible volcanic eruptions
5/5–9	W. Columbia	Powerful storms, deluges
5/8–17	NE Argentina	Major earthquake
5/12–14	Central Columbia	Powerful storms and winds
5/17–22	W. Columbia	Destructive storms, gales, floods
5/20–24	S. Cuba, Haiti	Destructive storms, winds, and/or powerful tremors
5/26–28	Costa Rica	Storms–could be powerful
6/9–12	E. Brazil, N. Bahia	Storms
6/14–24	E. Brazil, N. Bahia	Storms, deluges, major floods
6/16–21	S. Peru	Deluges, floods
6/16–21	Lesser Antilles, Windward Islands	Storms, deluges, floods
7/1–14	N. Brazil, Guyana	Severe storms, major floods
8/4–12	N. Columbia	Fierce storms, gales, major floods
8/5–14	E. tip of Brazil	Fierce storms, gales, major floods
8/11–16	E. Venezuela	Powerful storms, deluges, floods
8/14–18	Lesser Antilles N. Windward Is.	Storms
8/16–24	W Caribbean, Cuba	Hurricane
8/22–25	Mato Grosso, Bahia	Storms
8/26–9/3	Yucatan, Honduras, Guatemala	Hurricane
8/26–30	Lesser Antilles, N. Windward Is.	Fierce storms, powerful winds
8/26–30	SW to SE Brazil	Fierce storms, powerful winds
8/29–31	N.Chile coast	Fierce storms, powerful winds

Dates	Locations	Events
9/13–20	Panama, Columbia	Fierce storms, winds
9/13–21	E. Brazil, Maranhao	Fierce storms, major floods, chance of tremors
9/15–25	N. Chile coast	Fierce storms, powerful winds
9/16–24	NE Bolivia	Fierce storms, powerful winds, or powerful tremors
9/21–25	Lesser Antilles, N. Windward Is.	Fierce storms, powerful winds
9/21–25	E. Brazil	Storms, deluges
9/27–10/6	Suriname	Powerful tremors
9/29–10/5	NE Mato Grosso	Deluges, floods
9/29–10/5	Nicaragua, Costa Rica	Deluges, floods
10/2–10	SE Venezuela, N. Brazil	Fierce destructive storms or powerful tremors
10/2–10	W. Brazil near Bolivian border, SE tip Brazil	Fierce destructive storms or powerful tremors; could be offshore
10/2–10	NE tip Brazil	Fierce storms, destructive winds
10/2–10	NE Columbia	Fierce storms, destructive winds, major floods
10/8–23	SE Mato Grosso	Deluges, floods
10/15–24	Lesser Antilles, N. Windward Is.	Fierce storms, powerful winds, possible volcanic activity
10/16–22	SE Brazil coast	Storms, deluges, floods
10/16–24	NE Bolivia	Powerful storms, deluges, floods
10/20–26	Nicaragua, Costa Rica	Torrential rain
10/20–25	Yucatan, Guatemala	Powerful storms, deluges
10/27–31	Honduras, Yucatan	Storms

Dates	Locations	Events
11/6–15	Yucatan, Honduras, Guatemala, Nicaragua, Panama	Powerful storms, deluges, floods
11/11–17	Central Argentina	Deluges, floods
11/18–24	Panama	Powerful storms, deluges, floods
11/18–24	E. tip Brazil	Powerful storms, deluges, floods
11/22–25	Suriname	Storms
11/24–27	E. Brazil, SE Para	Storms, deluges
11/27–29	N. Para, Territorio do Amapa	Storms
12/1–4	Central Brazil	Storms, deluges, floods
12/24–28	SW Columbia	Storms
12/25–29	Lesser Antilles, Windward Is.	Storms
12/27–31	Yucatan, Guatemala	Fierce storms, deluges, floods
12/28–31	E. tip Brazil, NE Venezuela	Fierce storms, deluges, floods

Africa

Dates	Locations	Events
1/14–20	E. Chad, W. Sudan, E. Cent. African Rep.	Fierce storms and deluges
1/20–23	NE Sudan, NW Ethiopia	Fierce storms, powerful winds, deluges, floods
1/20–27	SW Sudan, N. Zaire	Torrential rain, floods
1/25–30	NE South Africa	Deluges, floods, possible storms
1/26–31	Central Algeria	Torrential rain
2/4–10	Central Algeria, cent. South Africa	Storms

Dates	Locations	Events
2/5–8	Mauritania, Western Sahara	Storms
2/12–14	Cent. South Africa, E. Zaire, W. Uganda	Storms, deluges
2/21–25	SW Mali, Botswana, S. Mauritania	Storms, deluges, floods
2/22–3/2	Morocco	Powerful storms, deluges, floods
2/27–3/2	Sudan, Ethiopia	Storms, deluges
2/28–3/2	Central South Africa	Storms, deluges
2/28–3/2	E. Mauritania, W. Mali	Powerful storms, deluges, floods
3/10–24	NE Algeria	Major earthquake
3/14–26	Sudan	Deluges, floods and storms
3/15–28	E. Algeria, W. Libya	Deluges, possible floods
3/20–25	W. Madagascar	Fierce storms, deluges, floods, chance of seismic activity
3/26–30	SW Niger, N. Benin	Storms, torrential rain
3/26–31	South Africa	Powerful storms, deluges
3/28–4/2	NE Namibia, S. Angola	Powerful storms, deluges, floods
3/30–4/2	SE Zimbabwe	Powerful storms, deluges, floods
4/4–14	Sudan	Powerful storms, deluges, floods
4/5–10	E. Zaire	Major floods or major tremors
4/13–20	South Africa	Fierce storms, deluges, floods, or powerful tremors
4/16–20	Botswana	Powerful storms, deluges, floods
4/16–22	S. Mozambique, Algeria	Fierce storms, deluges, floods, or powerful tremors
4/18–22	N. Sudan, Namibia, E. Niger, W. Chad	Powerful storms, fierce winds, floods, or major tremors

Dates	Locations	Events
5/3–10	NW Sudan, SW Egypt	Powerful storms, deluges, floods
5/8–17	E. Niger, W. Chad, SW. Mali, S. Mauritania	Major quakes or powerful storms, major flooding
5/10–13	Ivory Coast, Ghana, Burkina Faso	Powerful storms, winds
5/17–22	Ivory Coast, Ghana, Burkina Faso	Fierce storms, gales, floods
5/17–23	NW Sudan, SW Egypt	Destructive storms, gales, floods
5/20–25	Central African Republic, S. Chad	Fierce storms, gales, floods
5/26–6/4	Libya, S. Algeria	Fierce storms, winds, floods
6/9–12	Egypt	Torrential rain
6/9–16	Libya	Powerful storms, deluges, floods
6/16–19	Botswana	Torrential rain
6/18–26	S. Libya, N. Chad	Powerful storms, deluges, floods
6/19–28	Mozambique, S. Zimbabwe	Powerful storms, deluges, major floods
6/25–7/16	Mauritania, Senegal, Guinea, Sierre Leone, Western Sahara	Hurricane, deluges, major floods
7/22–30	Mauritania	Storms, deluges, floods
8/2–12	NW Chad	Fierce storms, destructive winds, massive deluges, major floods
8/2–12	E. Mauritania, N central Mali	Fierce storms, destructive winds, massive deluges, major floods
8/14–18	Mauritania	Storms, could be powerful
8/15–18	E. Mali, Namibia	Powerful storms
8/21–24	Mauritania, Namibia	Powerful storms, deluges

Dates	Locations	Events
8/26–31	E. Mali, Namibia, Mauritania, Western Sahara	Destructive storms and winds
9/13–20	Coastal Somalia	Tropical cyclone
9/15–23	NE Sudan, SE Egypt, E. Algeria	Powerful quakes or fierce storms, fierce winds, deluges, floods
9/17–25	Niger	Destructive storm or powerful quake
9/22–27	Mauritania, Senegal, Guinea	Storms
9/28–10/10	Uganda	Violent storms and winds
10/2–10	Mauritania, Senegal, Guinea	Extremely dangerous storms, could bring major earthquakes
10/2–10	Madagascar	Extremely dangerous cyclone
10/2–10	N. Namibia, S. coastal Nigeria	Extremely dangerous storms, could bring powerful tremors
10/5–10	Senegal	Extremely dangerous storms
10/13–18	N. Niger	Powerful storms
10/17–22	NW Atlantic coast	Storms, deluges
10/17–24	N. Mali, S. Algeria	Fierce storms, winds, deluges
10/20–23	Mauritania	Deluges
10/20–28	Libya	Fierce storms, powerful winds
10/27–30	S. Niger	Fierce storms, powerful winds
11/4–6	W. Egypt	Storm activity
11/6–14	NW Mali, Nigeria	Storms
11/11–17	N. Chad	Torrential rain
11/18–25	Western Sahara, Mauritania	Storms
11/22–29	NE Sudan	Storms
11/25–29	SW Ethiopia	Deluges, possible storms

Dates	Locations	Events
11/25–29	W. Mali	Storms, deluges
12/1–3	Central Sudan	Deluges, possible storms
12/5–8	S. Algeria, SW Libya	Deluges
12/18–24	Central Mauritania	Powerful storms, winds, deluges
12/25–28	Coastal Cameroon	Storms, deluges
12/26–31	Egypt, N. Sudan	Powerful storms, deluges, floods

Chapter Five

A Stormy Start: 2001

As the year 2000—one of the worst years ever recorded for natural disasters—finally draws to a close, we complete a very eventful century in which humanity has made great advances in space exploration, science, medicine, and technology.

The year 2001 heralds the beginning of the new millennium, and I'm happy to say the planetary patterns for 2001 indicate a less active year for severe earthquakes and volcanic activity than we saw in 2000. Not that there won't be any major seismic activity to be concerned about, but it generally will not be as frequent nor quite as severe as that of the previous year. That is, with the exception of the mid-June period and the period from mid-September to early October when major quake patterns will be strongest.

Weather patterns for the year 2001 will be severe at times. We are likely to see a round of major floods in the periods of February, late September through October, and especially in May through July.

Fierce Weather and Seismic Activity, May–June 2001

While seismic activity may be lessened in the year 2001, the area of major concern is the fierce weather patterns targeting various locations around the world in the months of May and June—particularly June. The weather patterns will be fierce from June 10–21. The solstice on June 21 will be followed hours later by the solar eclipse at zero degrees of Cancer, which will be quite significant.The planetary influences around June 10–21, and especially the last quarter moon phase on June 14, could trigger some of the most significant events to occur in 2001.

The powerful planetary influences in play from June 10–21 could trigger a major tremor in California south of San Francisco, Atlantic and Gulf of Mexico hurricanes, and a powerful tropical cyclone or major seaquake in the Coral Sea that could threaten the coastline of Queensland, Australia.

The danger of seaquakes, particularly in the Pacific Ocean, is that they can generate powerful tsunami that can sweep across the ocean. In deep waters they are hardly detectable at all, appearing more like a ripple on the ocean where the crest-to-trough measurement is usually no more than a few feet high. Upon reaching shallow waters along some unsuspecting shoreline, however, the waves begin to slow down and build into towering walls of water, where they sweep coastlines clean of all life and vegetation. Although tsunami strike in a series of waves, they do not follow any set order or sequence. Sometimes smaller twenty-foot waves will strike before the "big one" hits, or the big wave will strike first followed by a series of smaller waves. Other times a series of large waves of equal size can continue to pound the shoreline for hours.

On July 17, 1998, a powerful quake struck in the southwest Pacific Ocean just north of Papua New Guinea. It generated a powerful tsunami which moved south to target the unsuspecting nearby island communities, devastating everything in its path and sweeping well over seven thousand people out to sea.

Location charts indicate that some event will definitely take place in the eastern Coral Sea southwest of the Solomon Islands or northwest of New Caledonia around June 12–21. This could be either a seaquake or a very late-season tropical cyclone. Exactly what that event will be I cannot say for sure, but if it does turn out to be a tropical cyclone, this system will be really fierce and possibly generate powerful storm surges as it moves toward the Queensland coastline. In that event, there will be ample warning for Queensland. However, should a seaquake strike in the region of the eastern Coral Sea, quake-generated tsunami would strike Queensland's shores within hours, leaving very little time for evacuation procedures.

That the fierce, destructive June 12–21 aspects are affecting the Coral Sea at a time when ocean temperatures are cooler and atmospheric conditions are more stable and not conducive to tropical cyclone activity makes it difficult to determine exactly what will happen. The same aspects influencing the Coral Sea during the summer months, when the tropical cyclone season is in full swing, would produce a category five tropical cyclone the size of Agnes (300 miles across), or Joy, which targeted the north Queensland coast in December 1991.

Australia's Queensland Coastline Affected June 14–21

Although tropical cyclone activity in the Coral Sea in June is very rare, it has happened before. In 100 years of tropical cyclones forming in Australia's Coral Sea region, four tropical cyclones have formed in the eastern Coral Sea in June and headed for the southeast Queensland coast. Because of cooler ocean waters, these systems dissipated at sea and never made landfall.

The intensity of this system will very much depend on just how warm the ocean temperatures will be in mid-June. It is quite a possibility that in 2001 we may see a return of El Niño. In that case, warmer ocean waters may well favor late tropical cyclone development.

If a very powerful tropical cyclone does form in the eastern Coral Sea region around the Solomon Islands or New Caledonia around June 13–21, it will be headed toward the Queensland coastline. Powerful winds from this fierce storm could whip up ocean waves driving a storm surge ahead of the storm and send them crashing onto Queensland's beaches.

If a tropical cyclone does not form at this time, chances are this planetary influence could trigger a major seaquake caused by stresses along the Indian-Australian/Pacific tectonic plates. This could trigger a powerful tsunami headed toward the Queensland coast. However rare, a late tropical cyclone is far more probable than a seaquake.

The Great Barrier Reef, which spans almost the entire length of Queensland on Australia's east coast, generally serves as a protector against storm surges generated by tropical cyclone activity in the Coral Sea. However, on

March 5, 1899, a fierce tropical cyclone, (termed a hurricane due to its intensity) whipped up sea waves sending forty-foot storm surges pounding into Bathurst Bay in far north Queensland. According to a report handed to the Brisbane Bureau of Meteorology by H. E. Whittingham in July 1958, the huge waves engulfed an unsuspecting pearling fleet anchored in Bathurst Bay and swept three miles inland along the remote largely nonpopulated region of far north Queensland. The entire pearling fleet, comprised of eight-six vessels, sank beneath the waves. All boats recovered close to shore had their masts missing, having been rolled over and over at the bottom of the sea by the tremendous force of the waves.

Location charts for Queensland show two potential strike areas around this time period. One is far north Queensland and the other is southeast Queensland, namely Brisbane to Surfers' Paradise.

Early Hurricanes in the Gulf of Mexico and Atlantic Ocean

Although the U.S. Atlantic hurricane season officially begins in June, the Atlantic ocean temperatures are usually a little cool then for a system to fully develop into a fully mature hurricane. If the same planetary aspects were to occur during August or September at the height of the Atlantic hurricane season, it would produce a fierce category four hurricane the size of Andrew which slammed into Florida in August 1992, or Hurricane Hugo which devastated Charleston, South Carolina in September 1989.

However, the June 12–21 planetary influences are so strong that they have the potential to produce powerful hurricanes even this early in the year. The Atlantic storm may move west to influence the far northeast coast near New England, but should not be a severe as the storm in the Gulf which could target Mexico's Gulf coast or the Yucatan peninsula. The tropical storm forming in the Gulf around June 10–18 could well intensify into a fully fledged, very dangerous hurricane that creates havoc wherever it makes landfall. This system is coming with a deluge and will cause major flooding similar to what Honduras endured after hurricane Mitch ravaged the country.

Before this, though, a tropical depression could form in the Gulf of Mexico's Bay of Campeche around May 23–31. Although less severe, this system could have powerful winds, and also cause some major floods around Mexico's Gulf States.

Powerful Seismic Activity in June

Not only will fierce weather dominate weather patterns in May and especially June, but powerful earthquakes will occur as well. The most critical time period for major tremors to occur will be around the June 10–21 period. Although fierce storms and floods may be experienced in California around the June 14–21 period, the potential is also great for California to sustain the brunt of a powerful earthquake. Regions from San Francisco to Monterey could be affected around June 14–18, and southern California to the Nevada/Arizona border around June 19–28

Further east in the Midwest lies the New Madrid Fault centered in Missouri, Kentucky, and Tennessee. Apart from major floods which can be expected to target both Kentucky and Tennessee in the May–June period, one of these regions, or the New Madrid fault itself, could experience powerful tremors around June 12–18.

From June 12–21 Okinawa and southwest Japan—particularly Honshu, Kyushu, or Shikoku—could be hit by powerful tremors, and a typhoon as well.

Floods in May, June, and Possibly July

India, ravaged by monsoon floods in 2000, will be largely spared in 2001, and although heavier-than-normal monsoon rains could cause extensive flooding in the month of July, India will escape the large-scale flooding that Brazil and the United States are likely to experience. Monsoon floods can be expected to target Bangladesh, Assam, and Burma from June 21–30. Late monsoon floods will hit India and Bangladesh from August 20 to early September.

For the second year in row, heavy soaking rains could affect the eastern U.S.A. around May 24–June 20, plaguing Tennessee, Kentucky, Ohio, Mississippi, and possibly Louisiana. The Mississippi and Ohio rivers could overflow their banks, creating havoc and widespread floods.

Brazil could experience widespread floods throughout various states in May, June, and part of July. The same applies to Sumatra, Indonesia, Papua New Guinea, and New Britain,

Central to northeast Iran, Indonesia, England, and possibly Scotland could experience deluges and floods in June.

Wintery conditions for the southern hemisphere will begin in June with very heavy snowfall anticipated to fall in higher elevations. July also will bring good snowfall.

Major Earthquake Activity in 2001

Every year, hundred of earthquakes averaging 4.0 on the Richter scale are experienced around the world, so tremors are a normal part of life on earth. The bigger, more dangerous quakes do not strike all that often, and following a highly active year in 2000, 2001 should be far less active. Several locations that are most likely to be affected by powerful tremors in the year 2001 are listed below.

Period	Location
May	
5/1–31	Hawaii
5/18–31	Southern Italy
5/23–6/17	Indonesia (Sumatra), Papua New Guinea (New Britain)
June	
6/5–22	Coastal California
6/9–22	Iran
6/10–20	Tennessee, Kentucky, New Madrid Fault
6/14–21	Brazil (Para or Santa Carina)
6/15–20	Japan (Kyushu or Shikoku)
6/27–7/5	California
September	
9/29–10/6	Northeast Mexico
October	
10/1–8	Northern Italy

Period	Location
10/1–8	Japan (Kyushu and southern Honshu)
10/10–16	Brazil (Amazonas and Matto Grosso)
10/12–27	Japan (Kyushu and southern Honshu)
10/14–27	Northern Italy

The Most Violent Storms of 2001

You should pay attention to the following dates for major storms if a storm is brewing in your local area around the following dates. Whether or not the local weather reports are warning of rapidly worsening conditions, be aware that the storm could be a dangerous one and take precautions to assure your safety.

Period	Weather Event
1/1–5	Fierce storms, blizzards, heavy snowfall, tornadoes, hail, floods
1/17– 2/6	Fierce storms, blizzards, heavy snowfall, ice storms, gale-force winds, tornadoes, hail, floods, tropical cyclones
2/8–14	Fierce storms, blizzards, heavy snowfall, ice storms, gale-force winds, tornadoes, hail, deluges, tropical cyclones
2/15–24	Fierce storms, heavy snowfall, fierce winds, tornadoes, hail, major floods, tropical cyclones
3/8–18	Powerful storms, heavy snowfall, powerful winds, tornadoes, floods, tropical cyclones
3/25–4/3	Fierce storms, heavy snowfall, gale-force winds, floods, tornadoes, tropical cyclones

Period	Weather Event
4/13–15	Powerful storms, powerful winds
4/25–5/2	Powerful storms, powerful winds, tornadoes, deluges
5/5–10	Fierce storms, gale force winds, tornadoes, hail, floods
5/15–6/2	Destructive storms, fierce winds, tornadoes, hail, floods, tropical storms
6/7–10	Destructive storms, fierce winds, tornadoes, floods, tropical storms
6/12–21	Potentially the most destructive storms of the year, hurricane-force winds, destructive tornadoes, major floods, hail and lightning, fierce tropical storms, tsunami, storm surges
6/25–7/3	Powerful storms, powerful winds, major floods, tornadoes, tropical storms
7/11–15	Powerful storms, powerful winds, major floods, tropical storms
7/19–21	Powerful storms, powerful winds, major floods, tropical storms/hurricanes, typhoons
7/25–27	Powerful storms, powerful winds, major floods, tropical storms/hurricanes, typhoons
8/22–31	Fierce storms, fierce winds, hail, tornadoes, powerful hurricanes, typhoons
9/7 –15	Fierce storms, fierce winds, floods, powerful hurricanes, typhoons
9/19–23	Powerful storms, strong winds, floods

Period	Weather Event
9/29–10/7	Fierce storms, destructive winds, hail, tornadoes, major floods, destructive hurricanes and typhoons
10/12–17	Fierce storms, destructive winds, hail, major floods, destructive hurricanes and typhoons
10/23–27	Fierce storms, destructive winds, tornadoes, hail, major floods, destructive hurricanes and typhoons
11/15–30	Fierce storms, early snowfall, destructive winds, hail, tornadoes, deluges, tropical cyclones
12/3–7	Powerful storms, powerful winds, early snowfall, floods, tropical cyclones
12/9–15	Powerful storms, powerful winds, early snowfall, floods, tropical cyclones
12/22–30	Fierce storms, fierce winds, heavy snowfall, possible tornadoes, floods, tropical cyclones
12/28–30	Powerful storms, powerful winds, heavy snowfall, possible tornadoes, floods

Chapter Six

Tropical Cyclones, Hurricanes, and Typhoons in 2001

Storms are constantly raging on the planet. At any one time there can be up to 1,800 storms occurring simultaneously all over the world. Most of these storms are fairly inconsequential and do relatively little damage. However, savage storms often hit with very little warning because meteorologists fail to predict them, and so are harshly criticized for not warning the public. Several years ago Boris Yeltsin wanted to sack all of Moscow's meteorologists for failing to predict the most savage storm ever to hit Moscow. When hurricane-force winds raked southeast England, Belgium, and Holland, tearing roofs off houses, uprooting thousands of trees, and doing billions of dollars worth of damage, Dutch meteorologists blamed English meteorologists for not seeing it coming, and vice versa.

In Australia, meteorologists were also blamed for failing to predict the worst hailstorm ever to hit Brisbane and two of the worst known hailstorms to hit Sydney, one as recent as April 1999.

These fierce storms that meteorologists do not see coming are indeed predictable—but not by conventional methods. Through the medium of long-range weather forecasting one can tell what lies ahead and when fierce storms can be expected, and this is where astrometeorology comes in. For instance, these sudden violent storms always form within one to two hours under lightning-quick Mars/Uranus aspects—a condition that can be foreseen even years in advance.

The storm systems for the year 2001 will be just as intense as in the year 2000, but will mostly likely target different areas. The dates listed at the end of the previous chapter for the year's most powerful storms include the likely effects that could be experienced in the northern hemisphere as well as the southern hemisphere. For example, plummeting temperatures, snow, and blizzard conditions pertain to those areas that normally experience snow and deep freezes during the winter months.

Tornado activity, mostly experienced in America's midwestern and southern states peaks around the months of April and May. Sometimes, however, even during winter months tornado activity in the South and sometimes the Midwest can be severe, as shown by Florida's devastating tornadoes that hit in February 1998. So listed tornado activity is mainly for the United States, but could also occur in other regions of the world that may be susceptible to this type of activity.

The winter months of May/July 2001 for the southern hemisphere could see heavy snowfall in higher elevations or mountainous regions, and even in low-lying regions, especially in June and July. The months of June and July for the northern hemisphere summer could see severe weather conditions including fierce storms systems, gale-force winds, and major floods. It is also likely that a powerful tsunami could be generated in the Pacific Ocean around this time period. Any fierce tropical storms that form in any of the world's oceans during June 14–21 have the potential of becoming category five systems accompanied by powerful storm surges.

The dates of August 22–31, September 11–23, September 29–October 8, and October 13–27 can also see very intense hurricanes and typhoons in the world's oceans as well as fierce storms on land accompanied by gale-force winds and floods.

Tropical Storm Schedule for 2001

The year 2001 could bring a return of El Niño and a further warming trend. If that should happen, hurricane development will be largely retarded in the Atlantic, Caribbean, Gulf of Mexico, and in the northeast Pacific off Mexico. Typhoon numbers in the northwest Pacific region should remain unchanged as the El Niño doesn't appear to reduce tropical storm activity in this part of the world. Neither does it appear to affect tropical cyclone formation in the Indian Ocean. Occurrences of tropical cyclones around the northern Australian region, however, could also be severely reduced should El Niño return. Without any intervention from El Niño, the following tables indicate where and when tropical storms are likely to form in the year 2001.

Tropical Cyclones: SW Pacific, Coral Sea, Arafura Sea, Timor Sea, Indian Ocean, Arabian Sea, Bay of Bengal

Period	Location
1/15–20	Arafura Sea, Gulf of Carpentaria
1/21–30	S. Indian Ocean
1/21–27	Coral Sea
1/24–30	SW Pacific, Coral Sea
2/8–16	SW Pacific
2/15–20	Arafura Sea, Gulf of Carpentaria
2/8–15	S. Indian Ocean
3/23–30	Arafura Sea, Gulf of Carpentaria
3/30–4/6	Coral Sea
4/25–5/3	S. Indian Ocean
5/25–6/5	E. Coral Sea
6/14–22	Coral Sea
6/14–24	Arabian Sea and Indian Ocean
11/15–21	E. Coral Sea
11/24–29	SW Pacific
12/22–30	Arafura Sea, Gulf of Carpentaria
12/22–30	W. Australia
12/30–31	S. Indian Ocean

Typhoons NW Pacific Ocean, East China Sea, South China Sea, Andaman Sea, Philippine Sea

Period	Location
2/15–20	W. Pacific, E. of the Philippines
5/23–6/20	NW Pacific, South China Sea, Gulf of Tongking

Period	Location
6/14–21	NW Pacific/East China Sea
6/21–30	N. cent. Pacific
7/3–13	South China Sea, northwest Pacific
7/21–29	NW Pacific South China Sea
8/21–31	N. cent. Pacific
9/29–10/8	NW Pacific, South China and East China Sea
10/11–25	NW Pacific, South China Sea

Hurricanes
NE and N. Central Pacific Ocean

Period	Location
5/25–6/3	NE Pacific
6/14–24	NE Pacific
7/17–25	NE and N. cent. Pacific near Hawaii
8/20–28	N.cent. and northeast Pacific off Mexico
8/26–31	NE Pacific off Mexico
9/9–16	NE Pacific off Mexico
10/2–6	NE Pacific off Mexico
10/12–17	N. cent. Pacific near Hawaii
10/22–28	NE Pacific off Mexico

Hurricanes
Atlantic Ocean, Caribbean Sea, Gulf of Mexico

Period	Location
5/25– 6/2	Gulf of Mexico, Bay of Campeche
6/10–19	Gulf of Mexico
6/14–24	Atlantic
7/18–27	Atlantic, E. Caribbean
8/20–29	Atlantic
8/23–31	Gulf of Mexico
9/12–20	W. Caribbean off Cuba

Period	Location
10/1–10	Atlantic
10/13–19	E. Caribbean
10/23–27	W. Caribbean off Cuba

CHAPTER SEVEN

Event Timetable: 2001

Europe

Dates	Locations	Events
1/1–5	Italy, SE Germany	Powerful storms, deluges
1/16–25	Spain, NE Portugal	Fierce storms
2/2–8	Spain	Powerful storms
2/8–10	France, Belgium, Holland	Storms
2/8–12	Ireland	Storms
2/12–16	SE England, English Channel	Powerful storms
2/14–20	Ireland, England	Fierce storms, deluges, or heavy snowfall
2/15–18	Spain	Fierce storms, deluges
2/18–22	Portugal, Spain	Fierce storms, deluges

Dates	Locations	Events
3/13–22	France, W. Germany	Fierce storms, powerful winds
3/27–31	France, W. Germany	Powerful storms, deluges
4/1–3	SE England	Powerful storms
4/7–9	Sweden	Storms
4/12–13	Austria, Czechoslovakia	Storms
4/15–19	Italy, Germany	Storms
4/21–22	W. coast Ireland	Storms
4/25–5/1	W. coast Ireland	Powerful offshore storms
5/8–11	France, England	Storms
5/10–13	S. Italy	Fierce storms, possible tremors
5/14–19	France, England	Fierce storms, powerful winds
5/18–21	S. Italy	Fierce storms, possible quake
5/22–28	W. coast Ireland	Powerful storms
5/27–31	France, England	Powerful storms
5/27–31	S. Italy	Fierce storms, powerful winds, floods, possible tremors
5/29–6/7	England	Fierce storms, powerful winds
6/6–8	W. coast Ireland	Storms, strong winds
6/14–21	England, Scotland	Fierce storms, powerful winds, deluges, floods
6/20–28	SE England, Belgium, N. France	Powerful storms, deluges
6/21–25	Morocco, Portugal	Powerful storms
6/28–30	SE France	Storms

Dates	Locations	Events
7/11–15	SE England, Belgium, N. France, possibly Norway	Powerful storms, deluges, floods
7/20–28	E. Sweden, Latvia, Estonia	Powerful storms, winds
8/10–12	Ireland	Powerful storms
8/11–12	S. Italy	Powerful storms
8/12–14	S. Spain	Storms
8/17–19	S. Spain	Storms
8/20–23	S. Italy	Fierce storms, powerful winds
8/20–25	Greece	Fierce storms, powerful winds
8/25–31	Morocco, Portugal	Fierce storms, powerful winds
8/27–31	Ireland	Fierce storms, powerful winds
9/10–15	England	Powerful storms, deluges, floods
9/17–21	Ireland	Storms
9/29–10/6	S. England, N. France, Belgium	Destructive storms, deluges, floods
10/1–8	N. Italy	Destructive storms, floods, or major tremor
10/10–17	Crete, SW Turkey	Destructive storms, possible tremors
10/19–29	N. Italy	Powerful storms, floods or more powerful tremors
10/19–27	E. Mediterranean, Crete	Fierce storms or powerful tremors
10/23–30	Balearic Islands, Sardinia, Greece, S. Turkey	Powerful storms, possible tremors

Dates	Locations	Events
10/23–31	Ireland, England, Portugal, Spain, N. Morocco	Powerful storms
11/3–10	France, Spain, Mallorca	Storms
11/14–19	E. Germany, Poland, Czechoslovakia	Fierce storms, powerful winds
11/21–29	Ireland, England	Powerful storms
11/23–30	Czechoslovakia, Poland	Fierce storms, powerful winds, deluges or early snowfall
11/25–29	W. coast Ireland, possibly Portugal	Powerful storms

Middle East, Turkey, W. Russia

Dates	Locations	Events
1/1–5	Turkey	Powerful storms, heavy snow
1/1–5	S. Saudi Arabia, Yemen	Powerful storms, deluges
1/17–28	SE Iran	Fierce storms, powerful winds
1/20–22	Yemen	Storms
1/20–28	Libya	Fierce storms and winds
2/8–16	Uzbekistan, Turkmeniya	Fierce storms, powerful winds
2/15–22	Saudi Arabia	Fierce storms, powerful winds
2/18–26	Oman, S. Iran, Persian Gulf	Fierce storms, powerful winds
2/23–26	East Syria, Iraq	Powerful storms
2/28–3/2	SW Saudi Arabia	Storms

Dates	Locations	Events
3/3–5	Turkmeniya	Storms
3/6–9	SW Saudi Arabia	Powerful storms
3/6–14	Red Sea near Yemen, Ethiopia	Powerful storms
3/7–12	Turkmeniya	Powerful storms
3/8–14	Gulf of Aden, Yemen, Saudi Arabia	Storms
3/13–19	Arabian Sea off Pakistan, Iran	Powerful storms
3/16–21	S. Yemen, Djibouti, N. Somalia	Fierce storms, powerful winds
3/20–23	SW Iran	Powerful storms
3/25–31	Persian Gulf, Yemen, Eritrea, S. Iran	Fierce storms, deluges, floods
3/27–4/2	S. Iran	Powerful storms, deluges, floods
3/27–31	S. Yemen, Djibouti N. Somalia	Fierce storms, deluges, floods
4/17–22	Central and SW Iran	Storms
4/23–26	NW Iran, SW Iraq	Storms
4/25–30	NE Iran	Fierce storms, powerful winds
4/27–30	SE Turkmeniya	Fierce storms, powerful winds
4/30–5/5	E. Egypt	Storms
5/2–3	SW Iran	Powerful storms
5/15–19	N. Persian Gulf	Storms
5/17–20	SW Iran	Powerful storms
5/23–30	E. Africa, Turkey	Fierce storms
5/24–30	SW Iran	Fierce storms, powerful winds

Dates	Locations	Events
5/29–6/2	N. Saudi Arabia	Fierce storms
5/29–6/5	Somalia, Yemen, S. Saudi Arabia	Fierce storms
6/9–22	Central to NE Iran	Severe storms, floods, possible earthquakes
7/9–11	N. Turkey, Black Sea	Storms
7/9–13	NE Egypt	Storms
7/13–15	E. Iran	Deluges
7/17–21	North Turkey, S. Russia	Powerful storms, deluges
7/17–21	NE Egypt	Storms
7/22–28	Syria, Lebanon, Jordan	Powerful storms
8/11–15	Central Turkey	Powerful storms
8/20–24	SE Iran, W. Pakistan	
8/24–27	Yemen	Storms, deluges
8/25–30	Afghanistan, Turkmeniya	Fierce storms, powerful winds
9/1–9	Egypt, Sinai	Powerful storms
9/10–12	Caucasus	Storms
9/17–21	NE Saudi Arabia, SW Iraq	Fierce storms, powerful winds
9/24–27	S. Iran, N. Oman	Powerful storms, destructive winds
10/1–6	N. Oman	Fierce storms, very destructive winds
10/2–6	N. Afghanistan	Fierce storms and winds
10/10–15	Central/NW Egypt, possibly Crete	Fierce storms, powerful winds

Dates	Locations	Events
10/12–17	S.Turkey	Fierce storms, powerful winds
10/16–31	E. Oman	Offshore storms, powerful winds
10/20–29	NE Egypt, Crete, possibly S. Turkey	Fierce storms, powerful winds
11/1–3	Yemen, S. Saudi Arabia	Powerful storms
11/7–9	Yemen, S. Saudi Arabia	Powerful storms
11/8–13	E. Iran, W. Pakistan	Storms, deluges
12/6–9	Yemen, Oman	Deluges
12/21–25	Yemen	Powerful storms, deluges
12/22–24	S. Iran	Storms
12/28–31	SW Iran, U.A.R., Eritrea	Fierce storms, destructive winds

Asia, India, Bangladesh, Indonesia, Philippines, Japan

Dates	Locations	Events
1/1–6	Palawan, Sulu Sea, Japan, Irian Jaya	Fierce storms, deluges
1/1–6	W. Mongolia, NW China, S. Burma, Malaysia	Fierce storms, deluges
1/9–13	W. China	Storms
1/16–18	N. Sumatra	Powerful storms
1/17–21	NW China, India	Destructive storms

Dates	Locations	Events
1/19–30	Thailand, Burma, E. Malaysia, Java, S. Sumatra, Borneo, Irian Jaya, Papua New Guinea	Destructive storms, powerful winds
1/19–30	S. China, Hainan Dao	Destructive storms, powerful winds
1/24–30	Nepal, SW China, NE India, Bangladesh	Destructive storms
2/1–6	E. Malaysia, S. Sumatra, W. Irian Jaya	Powerful storms, deluges, floods
2/1–20	W. coastal India	Fierce weather, storms, severe cold, possible floods
2/2–6	Cambodia, Laos, Vietnam	Powerful storms
2/3–20	Burma, Borneo, S. China Sea, N. Sumatra, Irian Jaya, Papua New Guinea, Solomon Islands	Fierce weather, powerful storms, major floods
2/8–15	Cent. Indian Ocean	Fierce tropical cyclone
2/14–20	W. India	Fierce weather, destructive storms, floods
2/15–24	W. Irian Jaya	Powerful storms, floods
2/18–24	Papua New Guinea	Storms
3/6–14	W. Irian Jaya	Powerful storms, destructive winds
3/7–10	S. Japan	Powerful storms
3/9–12	S. China	Storms

Dates	Locations	Events
3/9–15	S. India, Sri Lanka	Storms
3/13–20	Papua New Guinea, New Britain, Bougainville	Fierce, destructive storms, powerful winds
3/14–18	S. China	Destructive storms, powerful winds
3/16–20	Japan (Kyushu)	Storms
3/19–20	NW China	Storms
3/20–25	Papua New Guinea	Storms
3/24–4/1	Japan (Shikoku, SW Honshu)	Powerful storms
3/27–4/3	Papua New Guinea, New Britain	Destructive storms
4/1–3	Nepal, Tibet	Powerful storms
4/13–18	S. China, Assam, Bangladesh	Storms
4/20–21	N. Sumatra, Malaysia	Storms
4/23–26	W. Irian Jaya	Storms, deluges
4/25–5/3	Indian Ocean	Tropical cyclone
5/1–8	S. Philippines	Storms
5/7–10	India's east coast, Sri Lanka	Storms
5/13–19	Japan (Honshu), Irian Jaya	Storms
5/15–20	Malaysia, Sumatra	Destructive storms
5/15–18	Irian Jaya	Storms
5/16–19	N. Sumatra, Burma	Storms, deluges

Dates	Locations	Events
5/23–6/17	Malaysia, Indonesia (Sumatra), Papua New Guinea (New Britain)	Dangerous storms, floods, fierce typhoons, powerful winds and storm surges, possible tremors and volcanic eruptions
5/23–6/20	Vietnam, S. China, Laos, Cambodia, Hainan Dao	Dangerous storms, fierce winds, floods. Powerful typhoon in S. China Sea and Gulf of Tonkin
5/25–6/20	NW Pacific, S. China Sea, Gulf of Tonkin, Celebes Sea, Indian Ocean	Fierce typhoons, tropical cyclones, storm surges
5/29–6/8	Irian Jaya	Fierce storms, dangerous winds, floods
6/10–20	Irian Jaya	Severe weather, fierce storms, floods
6/14–20	Arabian Sea	Tropical cyclone
6/14–21	Okinawa, S. Japan	Fierce typhoon
6/21–24	Papua New Guinea	Fierce typhoon
6/21–28	E. Bangladesh, Burma	Powerful storms
7/10–15	S. India, SE Asia	Powerful storms
7/10–15	Philippines, Borneo	Powerful storms
7/13–20	Philippines, Moluccas	Powerful storms
7/15–19	Japan	Storms
7/16–22	SE Asia	Powerful storms
7/21–24	Japan	Powerful storms

Dates	Locations	Events
7/21–29	NW Pacific, S. China Sea, Sumatra, Cambodia, Vietnam	Typhoons
7/22–24	Irian Jaya, Papua New Guinea	Powerful storms
7/25–28	W. Borneo	Storms
8/4–7	Bangladesh, S. Assam	Storms, deluges
8/11–14	Celebes, Mindanao	Storms
8/19–24	Papua New Guinea, New Britain	Fierce storms, powerful winds, possible tremors
8/20–24	Celebes, Mindanao	Fierce storms, powerful winds
8/20–25	Bangladesh, S. Assam	Fierce storms, powerful winds
8/21–31	N. Pacific	Fierce typhoons
8/23–24	NE Pakistan, Kashmir	Powerful storms
8/24–31	Moluccas, W. Irian Jaya	Fierce storms, powerful destructive winds
8/27–31	NE India, Bangladesh, S. Assam	Fierce storms, powerful destructive winds, major floods
8/27–31	SE India, Sri Lanka	Fierce storms, powerful winds
9/2–9	Philippines, NE Celebes, Moluccas	Storms
9/10–12	NW Sumatra	Storms

Dates	Locations	Events
9/29–10/5	NW Sumatra	Fierce storms, floods
9/29–10/8	Japan	Destructive storms, typhoon, possible tremors
9/29–10/8	NW Pacific, S. China Sea, Hainan Dao	Fierce typhoons
10/1–4	NE India, W. Java	Powerful storms, deluges
10/1–6	Papua New Guinea, SE China	Fierce storms, powerful winds, floods
10/2–5	Timor, Indonesia	Powerful storms
10/10–14	Malaysia, Borneo, Sabah	Fierce storms, powerful winds
10/11–17	Japan, Philippines, Moluccas	Fierce storms, powerful winds, seismic activity
10/13–18	Japan	Fierce storms and winds
10/21–27	NE India, Malaysia, Indonesia, Sumatra, Java, Timor	Fierce storms, powerful winds
10/13–28	NW Pacific	Fierce typhoons
10/21–29	Papua New Guinea	Powerful storms, floods
10/21–27	S. China, east of Hong Kong	Fierce storms, possible typhoon strike
10/24–29	Celebes, Moluccas	Storms
10/24–29	Indonesia, Sunda Is.	Storms, deluges
11/7–10	Japan	Deluges
11/7–13	W. Irian Jaya	Deluges, floods
11/8–11	S. tip of India	Storms, deluges
11/14–21	Papua New Guinea, New Britain	Fierce storms, deluges, major floods

Dates	Locations	Events
11/15–23	S. tip of India	Fierce storms, deluges, floods
11/22–28	Bougainville	Fierce storms, deluges, floods
11/22–30	Sri Lanka	Fierce storms, winds, floods
11/30–12/2	NE Celebes, Philippines	Storms
12/3–8	Sumatra, Borneo, Irian Jaya, New Guinea	Powerful storms
12/9–14	Papua New Guinea, Irian Jaya	Storms
12/14–24	Indonesia, Celebes, E. Sunda Islands	Powerful storms, deluges, floods
12/21–31	Moluccas, Irian Jaya, Papua New Guinea	Fierce storms, powerful winds, floods
12/22–28	Borneo, Palawan, Celebes, Mindanao	Fierce storms, deluges, floods
12/22–25	Papua New Guinea	Storms, deluges
12/30–31	Indian Ocean	Tropical cyclone

Australia and New Zealand

Dates	Locations	Events
1/1–8	Arnhem Land, Northern Territory, Gulf of Carpentaria	Fierce storms, deluges, floods
1/11–16	South Australia, Victoria	Powerful storms
1/15–20	Gulf of Carpentaria, N. Queensland	Fierce storms, major floods
1/16–21	North Island N.Z.	Storms

Dates	Locations	Events
1/18–23	E. South Australia, W. Victoria	Fierce storms
1/20–30	Gulf, Coral Sea	Fierce tropical cyclones
1/24–30	N. and coastal Queensland	Fierce storms, major floods
1/21–24	E. South Australia	Destructive storms, deluges, floods
1/22–23	Victoria, N.S.W.	Destructive storms, powerful winds,floods
1/24–27	South Island N.Z.	Deluges, floods
1/27–2/6	North Island N.Z.	Powerful storms, floods
1/29–2/2	E. NSW	Powerful storms
2/1–6	Central/SE Queensland	Fierce storms, possible cyclone
2/1–14	SE Queensland NE NSW	Fierce storms, powerful winds, deluges, floods
2/1–8	S. Victoria	Powerful storms, floods
2/1–6	South Island N.Z.	Storms, deluges, floods
2/8–16	SW Pacific	Fierce tropical cyclone
2/10–20	South Island N.Z.	Destructive storms, floods
2/10–16	E. NSW, E. Victoria	Fierce, destructive storms, gale-force winds, deluges
2/15–20	Gulf of Carpentaria	Powerful tropical cyclone, major floods
2/18–24	Northern Territory near Alice Springs	Fierce storms, floods
2/23–27	Cape York	Storms, floods
3/6–14	Arafura Sea, Northern Territory, Arnhem Land	Powerful storms, floods, possible cyclone

Dates	Locations	Events
3/10–18	Coral Sea	Tropical cyclone moving east
3/15–20	Cape York, N. Queensland	Storms, deluges
3/16–20	Western Australia	Powerful storms
3/20–31	Northern Territory, Barkly Tableland	Powerful storms, deluges, floods
3/21–25	North Island N.Z.	Storms, deluges
3/25–4/3	Northern Territory	Powerful storms, deluges
3/29–4/3	Cape York, far N. Queensland	Powerful storms, floods
3/30–4/6	Coral Sea	Tropical depression, cyclone
4/8–12	NW Western Aust.	Storms
4/23–26	Gibson Desert, Arnhem Land	Storms, deluges
4/25–30	NE Queensland NE Western Aust.	Powerful storms, deluges
5/1–11	Central/SE Queensland	Powerful storms, floods
5/2–10	Great Sandy Desert	Powerful storms, deluges, floods
5/7–10	New Caledonia	Powerful storms
5/7–15	SE Queensland	Fierce storms, powerful winds
5/13–20	South Island N.Z.	Fierce storms, powerful winds, floods
5/13–20	Adelaide, Victoria	Powerful storms and winds
5/14–18	W. of North Island N.Z.	Powerful storms
5/15–19	Arnhem Land, Gulf region	Fierce storms, winds
5/16–20	Queensland, NE NSW	Powerful storms,winds, deluges

Dates	Locations	Events
5/23–30	South Island N.Z.	Powerful storms
5/24–30	N. Queensland, Cape York Peninsula	Fierce storms, gale-force winds, floods
5/24–6/2	SE Queensland, NE NSW	Fierce storms, gale force winds
5/25–6/5	Coral Sea, SE Queensland	Late tropical cyclone
5/26–28	NSW	Fierce storms, gale-force winds
5/26–6/30	NSW, Victoria	Heavy snowfall on Alps
5/27–6/2	S. Australia, SW Northern Territory	Storms
5/27–6/2	NE Western Aust., Northern Territory	Fierce storms, powerful winds, deluges
6/14–21	NE Western Aust., NW Northern Territory	Destructive storms
6/14–22	Coral Sea, Queensland coast	Fierce tropical cyclone with storm surges, or major seaquake with possible tsunami
6/21–27	N. Queensland	Powerful storms, deluges
6/21–28	New Zealand	Powerful storms
7/16–24	Northern Territory	Powerful storms, deluges
7/17–21	Northern Territory Arnhem Land, Gulf regions	Storms, deluges, floods
7/25–29	NE Western Aust., Arnhem Land.	Storms, deluges, floods
7/27–30	SE Queensland	Offshore storms
8/12–14	North Island N.Z.	Storms
8/20–25	South Island N.Z., possibly North Is.	Fierce storms, powerful winds, very cold

Dates	Locations	Events
8/19–22	Queensland	Storms
8/25–30	South Island N.Z.	Powerful storms
9/2–4	SE South Aust.	Storms
9/7–15	SE South Aust., SW Victoria	Storms, deluges
9/10–15	North Island N.Z.	Storms
9/13–18	S. Victoria	Storms, deluges
9/17–20	Northern Territory, NE South Aust.	Storms
9/17–21	SE Queensland, N. NSW	Storms
9/20–23	S. Victoria	Deluges
9/22–25	SW Western Aust.	Deluges
9/25–28	SE Queensland, N. NSW	Storms
10/2–4	SE Queensland	Powerful storms
10/2–6	Northern Territory, NW South Aust.	Powerful storms
10/2–7	New Caledonia, Vanuatu	Fierce storms
10/2–7	SE South Aust.	Fierce storms, floods
10/10–16	SW Victoria	Fierce storms, powerful winds
10/19–28	SW Victoria, SE South Aust.	Bad weather, storms, floods
10/23–28	NE Western Aust.	Fierce storms, deluges, floods
10/23–28	Vanuatu	Fierce storms, deluges, floods
10/23–29	Cape York, N. Queensland	Storms, deluges

Dates	Locations	Events
10/23–29	Gibson Desert, Great Victoria Desert	Deluges
10/24–29	North Island N.Z.	Fierce storms, powerful winds, floods
10/27–30	Northern Territory, Melville Island	Storms
11/5–11	Cape York, Queensland	Deluges
11/7–10	Arnhem Land	Storms and deluges
11/12–15	N. T. Gulf, NW Queensland	Powerful storms, deluges, floods
11/15–17	South Island N.Z.	Storms
11/15–21	Coral Sea, New Caledonia	Fierce storms, possible cyclone
11/15–25	Cape York, N. Queensland	Powerful storms, deluges, floods
11/24–27	South Island N.Z.	Fierce storms
11/24–29	SW Pacific, Fiji	Tropical cyclone
11/25–30	Cent. Queensland	Fierce storms, deluges, floods
11/30–12/4	Great Sandy and Gibson Deserts	Storms
12/4–7	Brunswick Bay to Cp. Londonderry	Powerful storms, deluges
12/7–10	Flinders Range S.A.	Powerful storms
12/7–10	Cape York, NW Queensland	Storms
12/9–20	NW Queensland	Storms, deluges, floods
12/10–16	Arafura Sea, Arnhem Land, Northern Territory	Storms

Dates	Locations	Events
12/22–30	E. Gulf, Cape York, N. Queensland	Powerful storms, major floods, tropical cyclone
12/22–30	J. B. Gulf to Arnhem Land	Powerful storms, major floods, possible tropical cyclone
12/30–31	Coastal NW Western Austalia	Powerful storms, floods

USA and Canada

Dates	Locations	Events
1/1–5	NW California	Powerful storms, deluges
1/9–12	Vancouver Is., Washington state	Storms
1/16–18	Florida, Georgia	Storms
1/20–30	Florida, Carolinas	Fierce storms, tornadoes
1/20–30	East Coast	Heavy snowfall, intense cold, blizzards possible ice storms
1/24–26	Michigan, Great Lakes, New England	Heavy snowfall
1/29–2/8	W. Tennessee, W. Kentucky, Mississippi, S. Illinois	Powerful storms, deluges, possible tornadoes
1/29–2/8	New York, S. Quebec, New England	Powerful storms, blizzards, heavy snowfall, possible ice storms
2/1–6	N. Arizona, S. Utah	Powerful storms, cold, deluges
2/2–8	New Brunswick, Nova Scotia	Powerful storms, heavy snowfall severe cold

Dates	Locations	Events
2/8–14	Indiana	Powerful storms, snow
2/8–14	S. New Mexico, upper Mexico	Powerful storms, plummeting temperatures
2/8–14	Maine, New Brunswick	Fierce blizzards, heavy snow
2/14–22	SE Ontario, Lake Huron	Powerful storms, heavy snowfall, blizzards
2/14–22	New Brunswick, Nova Scotia	Powerful storms, heavy snowfall, blizzards
2/14–22	Alabama, W. Carolinas, Pennsylvania, New York	Powerful storms, severe cold, heavy snowfall
2/14–23	NE Colorado, Nebraska	Powerful storms, very heavy snowfall
2/15–18	SW Yukon	Powerful storms, heavy snowfall
2/18–26	N. California	Powerful storms, floods
3/3–8	Vancouver Is., N. California	Storms
3/9–14	SE New Mexico, W. Texas	Storms
3/14–19	Idaho, Nevada, Utah, Arizona	Powerful storms, deluges
3/15–18	Central/E. Texas	Powerful storms, tornadoes
3/17–21	Montana, Wyoming, Colorado	Fierce storms, deluges
3/30–4/2	Louisiana	Fierce storms, deluges, tornadoes
3/21–24	NW California	Powerful storms, deluges
3/21–23	Maine	Storms
3/29–4/2	California	Powerful storms, floods

Dates	Locations	Events
4/6–8	Hawaii	Storms
4/8–10	W. Texas	Storms, possible tornadoes
4/13–17	Hawaii	Storms
4/25–5/1	Louisiana. Mississippi	Powerful storms, tornadoes
4/27–5/3–	Nevada	Powerful storms
4/27–5/3	NW California	Powerful storms, deluges
4/27–5/9	Hawaii	Powerful storms,deluges
5/1–3	W. Texas	Storms,
5/1–31	Hawaii	Powerful storms, or volcanic eruptions and tremors
5/6–9	Nevada, W. Utah, Arizona	Powerful storms, winds, deluges
5/6–9	Alabama	Powerful storms, winds, possible tornadoes
5/7–10	Oregon, California	Powerful storms, deluges
5/7–11	S. Dakota, Kansas, Nebraska, Oklahoma	Storms, powerful winds
5/12–20	Texas	Fierce storms, tornadoes
5/14–19	S. Dakota, Kansas, Nebraska, Oklahoma	Powerful storms, tornadoes
5/14–18	Montana, Wyoming, Colorado	Fierce storms and winds
5/15–18	Queen Charlotte Is.	Powerful storms
5/15–20	Washington, Oregon, California	Fierce storms, floods

Dates	Locations	Events
5/23–6/2	Mexico's Gulf states	Offshore tropical storm
5/24–31	Texas, Kansas, Oklahoma	Fierce storms, floods,tornadoes
5/24–6/7	S. California	Fierce storms, winds, floods, powerful tremor
5/24–6/20	Louisiana, Mississippi, Tennessee, Kentucky, Ohio	Many storms, floods, tornadoes
5/25–6/3	NE Pacific	Tropical storm/hurricane
5/29–6/3	Hawaii	Storms
6/5–13	N. California	Fierce storms
6/5–10	Hawaii	Powerful storms
6/5–17	Texas, Oklahoma	Fierce storms, floods, tornadoes
6/5–22	California coast	Destructive winds, massive floods, or powerful earthquake
6/10–18	Gulf of Mexico	Tropical storm
6/10–20	Louisiana, Mississippi, Alabama	Fierce storms, powerful winds, major floods
6/10–20	New Madrid fault, Tennessee, Kentucky	Powerful tremors
6/14–24	Atlantic, NE Pacific	Powerful hurricanes
6/21–27	British Columbia, Washington state, Oregon, California	Powerful storms
6/21–28	S. California, Lake Mead area	Powerful tremors

Dates	Locations	Events
6/23–26	Hawaii	Fierce storms and winds
6/25–28	Idaho, Colorado, New Mexico	Powerful storms
6/27–7/5	N. California	Tremors
7/5–11	Delaware, E. Pennsylvania, New York state	Storms
7/5–12	New England	Storms
7/9–12	Hawaii	Powerful storms
7/11–14	Nevada, Arizona	Tremors or storms and deluges
7/13–16	N. California	Storms, floods
7/15–21	NE Atlantic Coast	Powerful storms, deluges
7/16–21	Hawaii	Powerful storms
7/17–21	New England	Powerful storms
7/17–25	NE Pacific	Tropical storm or hurricane
7/18–27	Atlantic, E. Caribbean	Tropical storm–could intensify
7/20–23	Hawaii	Storms
7/22–24	SE Arizona, N. Mexico	Storms
7/24–28	N. California	Powerful storms, floods
7/25–28	New England	Powerful storms
7/25–28	Hawaii	Storms
8/4–6	S. Ontario, Michigan	Powerful storms
8/4–7	Mississippi, Alabama	Storms
8/10–12	N. Carolina	Powerful storms
8/12–14	Utah, Colorado	Storms

Dates	Locations	Events
8/19–27	Mexican W. coast	Offshore hurricane
8/19–22	Utah, Wyoming, Colorado, New Mexico	Powerful storms
8/20–24	S. Ontario, Michigan	Destructive storms
8/20–24	N. Carolina	Destructive storms
8/20–29	Atlantic Ocean	Powerful hurricane
8/21–23	S. Arizona, NW Mexico	Powerful storms
8/23–26	Louisiana, Arkansas, Missouri, Illinois	Powerful storms
8/23–31	Gulf of Mexico, Yucatan	Fierce hurricane
8/26–31	Mexican W. coast	Fierce offshore hurricane
8/27–31	Mississippi, Tennessee	Fierce storms, powerful winds
8/30–31	E. Texas	Powerful storms, tornadoes
9/9–16	Mexican W. coast	Hurricane
9/10–14	S. Florida	Powerful storms
9/12–20	W. Caribbean	Hurricane
9/22–24	Maine	Powerful storms
10/1–8	Minnesota, Iowa	Severe storms, deluges, floods
10/1–10	Atlantic	Fierce hurricane
10/1–10	S. Florida, Georgia, E. Cuba	Fierce storms, powerful winds, floods
10/2–6	NE Pacific off Baja	Hurricane
10/2–7	Nova Scotia	Fierce storms

Dates	Locations	Events
10/10–14	Hawaii	Fierce hurricane nearby
10/10–17	Illinois, Florida, Georgia	Fierce storms, powerful winds, floods
10/13–19	E. Caribbean	Powerful hurricane
10/21–28	British Columbia, Vancouver Is.	Fierce storms, powerful winds
10/22–27	Nova Scotia, St Lawrence Gulf	Fierce storms, deluges
10/23–27	W. Caribbean, Gulf of Mexico	Hurricane
10/24–27	Hawaii	Storms
11/1–3	New Mexico, Colorado	Storms
11/7–13	Texas	Storms, deluges, floods
11/8–10	N. Dakota, SE Saskatchewan	Storms, floods
11/13–19	Kansas, Oklahoma, Texas, Louisiana	Powerful storms, plummeting temperatures
11/14–19	Indiana, Kentucky Tennessee	Storms
11/20–27	Florida	Powerful storms
11/24–29	Louisiana	Powerful storms
11/25–28	Illinois	Storms
12/7–10	Oregon, Washington state, British Columbia, Alberta	Storms
12/8–12	N. California	Storms, deluges
12/14–16	S. California	Storms, deluges

Dates	Locations	Events
12/14–16	Queen Charlotte Is.	Powerful storms
12/21–22	S. Dakota	Storms
12/21–24	W. Alberta	Storms, heavy snowfall
12/22–24	Washington state, Oregon N. California	Storms
12/22–24	Nova Scotia	Powerful storms, intense cold, heavy snowfall
12/23–25	Oklahoma, Texas	Storms, deluges
12/23–25	S, Arizona, upper Mexico	Storms
12/30–31	Louisiana, Mississippi	Powerful storms, deluges, floods
12/30–31	Hawaii	Storms

Mexico

Dates	Locations	Events
1/1–5	Yucatan Peninsula	Powerful storms, floods
1/9–12	Baja	Storms
1/21–25	Baja	Powerful storms
2/1–5	Campeche, Yucatan, Baja	Powerful storms, deluges, floods
2/8–10	Oaxaca	Fierce storms, powerful winds
2/12–14	Guerrero	Storms, powerful winds, deluges
2/12–20	Jalisco, Yucatan Peninsula	Fierce storms, powerful winds, floods
2/23–25	Baja Peninsula	Storms
3/14–21	Tabasco, Yucatan, Campeche	Fierce storms, deluges, floods

Dates	Locations	Events
4/12–17	Campeche, Yucatan	Storms
4/22–27	Yucatan Peninsula, Guatemala, Belize	Storms, floods
4/25–30	Oaxaca, Tabasco, SE Vera Cruz	Fierce storms, powerful winds, floods
5/1–3	Tabasco, Chiapas	Storms
5/2–9	Tabasco, Chiapas	Powerful storms, floods
5/7–19	Offshore Baja	Powerful storms, fierce winds
5/14–20	Oaxaca, Tabasco, SE Vera Cruz, Chiapas	Fierce storms, powerful winds, floods
5/23–31	Tabasco, Chiapas Yucatan Peninsula, Vera Cruz	Powerful storms. Tropical storm in Bay of Campeche
5/29–6/3	Oaxaca, Puebla	Storms
6/1–23	Tabasco, Chiapas Yucatan Peninsula, Vera Cruz	Severe storms, major floods
6/13–20	Gulf of Mexico	Tropical storm/hurricane
6/21–27	Central/S. Mexico	Storms
7/9–14	S. Mexico	Storms
7/9–14	Baja	Powerful storms
7/17–24	Baja	Powerful storms, possible tremor
7/25–28	Baja	Powerful storms
8/4–7	Yucatan, Guatemala, Honduras	Storms, deluges
8/20–24	Jalisco	Fierce storms, powerful winds
8/20–25	Baja, NW Mexico	Powerful storms

Dates	Locations	Events
8/23–30	Chiapas, Tabasco, Campeche, Yucatan	Powerful offshore hurricane
8/27–31	NE Pacific, Jalisco	Fierce hurricane
9/9–16	Baja, NW Mexico	Fierce storms
9/9–16	NE Pacific	Hurricane
9/29–10/6	E. Coahuila, Monterrey, Nuevo Leon	Destructive storms or major earthquake
10/1–7	SE Vera Cruz, Tabasco, Oaxaca	Powerful storms, deluges, floods
10/1–5	SE Vera Cruz	Powerful storms, floods
10/2–10	Baja	Offshore hurricane
10/12–17	Nayarit	Fierce storms, powerful winds.
10/17–18	E. Coahuila, Monterrey, Nuevo Leon	Storms
10/21–27	SE Vera Cruz, Tabasco, Oaxaca	Fierce storms
10/22–29	E. Coahuila, Monterrey, Nuevo Leon	Powerful storms, deluges, floods
10/22–28	S. Baja	Offshore hurricane, fierce storms, floods
10/24–28	W. to S. Mexico	Powerful storms
11/7–10	S. Oaxaca	Deluges, floods
11/13–16	Yucatan, Belize, Guatemala	Powerful storms, deluges
11/15–21	Oaxaca, Chiapas	Fierce storms, floods
11/19–26	Sonora	Fierce storms, floods

Dates	Locations	Events
11/23–29	Chiapas, Guatemala	Fierce storms, floods
11/25–30	Guerrero	Fierce storms, winds, floods
11/25–30	Chihuahua, Jalisco, Michoagan	Fierce storms, floods
12/22–24	Sonora, Oaxaca	Storms
12/30–31	Yucatan Peninsula	Fierce storms, floods

South America

Dates	Locations	Events
1/1–5	NE Columbia, NW Peru, Ecuador, E. Brazil	Powerful storms
1/3–10	SE Brazil	Storms
1/10–20	Peru	Severe cold, powerful storms, blizzards, heavy snowfall
1/13–16	NE Columbia	Storms
1/16–22	E. Brazil, N. Peru, Ecuador	Fierce storms
1/20–26	Panama, Cuba	Fierce storms, floods
1/20–27	Bolivia	Fierce winter storms, heavy snowfall in high elevations
1/25–30	Peru	Fierce storms, intense cold, blizzards, heavy snowfall
1/25–30	Columbia	Fierce storms, powerful winds, floods
2/1–10	Peru, Chile	Powerful storms, heavy snowfall, deluges

Dates	Locations	Events
2/2–8	Columbia, Venezuela	Powerful storms, floods
2/10–18	Venezuela	Fierce storms, powerful winds, major floods
2/10–20	Argentina, Bolivia	Fierce storms, powerful winds, heavy snowfall or floods
2/14–22	W. Brazil, Bolivia, Argentina	Powerful storms, deluges, floods
2/15–20	N. Brazil	Powerful storms
2/15–20	NE Para	Flooding
2/23–25	Pernambuco, Paraiba	Storms
3/6–13	NE Brazil	Powerful storms, winds, deluges
3/15–19	Central Brazil	Powerful storms, winds
3/20–22	N. Chile	Storms
3/23–28	Far E. Brazil	Powerful storms
3/25–28	Central Brazil	Powerful storms
3/25–30	E. Brazil	Fierce storms, floods
3/29–4/3	N. Brazil	Fierce storms, floods
4/15–17	Peru, E. Argentina	Storms
4/23–27	Maranhao, Ceara	Deluges
4/25–5/3	NW Para, E. Columbia	Powerful storms, deluges
4/30–5/2	SW Peru, Bolivia	Deluges
5/1–8	E. Venezuela, Guyana, Surinam N. Brazil	Powerful storms, deluges
5/8–10	E. central Brazil	Storms, deluges

Dates	Locations	Events
5/14–19	SE Brazil	Fierce storms, winds, floods
5/19–24	Panama	Fierce storms, powerful winds
5/25–30	Columbia	Fierce storms, winds, floods
5/27–6/2	Far E. Brazil	Powerful storms, deluges
5/25–30	S. Guyana, N. Brazil	Fierce storms, major floods
6/13–20	Brazil, S. Guyana	Destructive storms, major floods
6/14–21	S. Brazil	Earthquake, possibly major
6/21–28	S. Chile, S. Argentina	Fierce storms, powerful winds, floods
7/9–30	Peru, Columbia, Ecuador	Powerful storms, deluges, floods
7/17–27	N. Brazil	Powerful storms, deluges, floods
7/20–27	SE Columbia, Venezuela	Powerful storms, deluges, floods
7/21–26	N. Brazil	Powerful storms
8/10–13	Columbia, Peru, Ecuador	Storms
8/11–15	Guyana, N. Brazil	Powerful storms
8/20–25	Guyana, N. Brazil	Severe storms, powerful winds
8/25–28	E. Bolivia, French Guiana, E. Brazil	Storms
8/26–31	Honduras	Hurricane
8/27–31	N. Brazil	Severe storms, winds, deluges
9/2–10	Suriname/Brazil border	Powerful storms
9/10–15	Far E. Brazil	Powerful storms

Dates	Locations	Events
9/10–13	Cuba, Honduras, Panama	Powerful storms
9/17–20	E. Brazil	Powerful storms
9/17–19	S. Brazil	Storms
9/29–10/4	Columbia	Storms
10/1–8	S. Brazil	Fierce storms
10/1–8	E. Venezuela	Fierce storms
10/2–7	French Guiana	Powerful storms, deluges
10/10–17	Far E. Brazil	Fierce storms, powerful winds
10/20–25	Columbia, French Guiana	Fierce storms, destructive winds
10/10–16	W. Brazil	Destructive storms, deluges, floods. Major tremor possible
10/16–29	W. Brazil	Powerful storms, major floods
10/19–27	E. Venezuela, W. Brazil, Bolivia NE Paraguay	Fierce storms, deluges, floods
10/23–28	Honduras, Nicaragua	Offshore hurricane
10/25–30	French Guiana	Storms, floods
10/28–31	Far E. Brazil	Powerful storms
11/6–9	French Guiana	Storms, floods
11/7–10	N. Brazil	Deluges
11/24–30	Guatemala, Honduras	Fierce storms, deluges
11/30–12/2	N. Brazil	Storms
12/9–13	N. Brazil	Storms, deluges
12/12–16	E. Brazil	Storms, floods
12/14–23	Guyana, N. Brazil	Storms, floods

Dates	Locations	Events
12/21–28	Panama, far E. Brazil, Bolivia, Venezuela	Powerful storms, winds, floods
12/27–31	N. Columbia, central Brazil	Fierce storms, powerful winds, floods

Africa

Dates	Locations	Events
1/1–5	Libya, Nigeria, Ethiopia, Sudan, Somalia, Niger	Powerful storms, deluges
1/2–5	Yemen	Powerful storms, deluges
1/9–12	Senegal	Storms, deluges
1/18–27	Namibia	Fierce storms, winds, deluges
1/18–31	Senegal, Chad, Mauritania	Fierce storms, winds, deluges
1/24–27	Western Sahara	Storms
1/25–2/6	NW Mali, Angola, Zaire	Fierce storms, floods
2/2–6	Morocco, Mauritania	Powerful storms, floods
2/8–14	Sierra Leone, Guinea	Powerful storms, winds
2/8–15	S. Sudan, Zambia	Fierce storms, powerful winds
2/8–28	Saharan Algeria, NE Mali	Fierce sand storms
2/14–20	NE Mauritania, NW Mali	Powerful storms, major floods

Dates	Locations	Events
2/15–22	Saudi Arabia, Iraq	Fierce storms, powerful winds, deluges, possible floods
2/15–24	S. Sudan	Powerful storms, major floods
3/7–13	Ethiopia, Kenya	Powerful storms, strong winds
3/14–18	Nigeria	Fierce storms, powerful winds
3/16–22	Djibouti, Eritrea, Yemen	Fierce storms, powerful winds
3/20–22	S. Algeria, N. Niger	Storms
3/25–31	Nigeria, U.A.R.	Storms, deluges
3/29–31	Djibouti, Eritrea, Yemen	Powerful storms
4/15–18	Namibia, N. Niger, SE Algeria	Storms
4/22–30	N. Tanzania	Fierce storms, winds, deluges
4/25–5/7	Cent. African Rep.	Powerful storms, floods
5/2–8	W. Tanzania	Storms
5/13–16	Algeria, Yemen, Oman, N. Mali	Fierce storms, powerful winds
5/15–18	NE Somalia	Fierce storms, powerful winds
5/16–31	NW Namibia, W. Botswana, Algeria, Libya, N. Niger	Fierce storms, powerful winds, floods
5/24–31	Mozambique, Zimbabwe, Zambia	Fierce storms
5/24–31	W. Africa coast	Fierce offshore storms
5/29–31	Yemen, Oman, N. Mali	Powerful storms
6/13–21	NE South Africa, SE Kenya	Fierce storms, destructive winds

Dates	Locations	Events
6/14–21	Madagascar	Fierce storms, destructive winds, possible offshore cyclone
6/14–21	W. Africa coast	Offshore storms or hurricane
6/21–22	NE Mali, S. Algeria	Storms
6/26–19	NE Mali, S. Algeria	Powerful storms
7/6–12	W. Mauritania, Western Sahara, NE Nigeria, Cameroon	Powerful storms, deluges
7/15–21	W. Mauritania, Western Sahara, NE Nigeria, Cameroon	Powerful storms, deluges
7/16–20	Mozambique	Offshore storms
7/22–25	Guinea-Bissau, Senegal, Mauritania	Powerful storms, deluges
7/25–28	Mozambique	Offshore storms
7/25–28	W. Mauritania, Western Sahara, NE Nigeria, Cameroon	Powerful storms, deluges
7/25–28	Namibia	Storms
8/4–8	W. Africa	Heavy rainfall
8/11–14	Lower Egypt, NE Sudan	Storms
8/12–15	N. Mali, S. Algeria	Storms
8/20–30	Saudi Arabia	Fierce storms, powerful winds
8/23–30	S. Algeria, Ivory Coast, Ghana	Fierce storms, powerful winds

Dates	Locations	Events
8/25–30	Morocco, Mauritania	Fierce storms, winds
9/1–9	E. Uganda, Kenya	Storms
9/17–20	S. Nigeria, S. Iraq, Kuwait	Storms
9/22–24	E. South Africa	Offshore storms
9/29–10/6	S. Mozambique	Destructive storms, gales.
10/1–7	S. Nigeria, UAR, S. Iran	Destructive storms, gales, floods
10/2–5	S. Malawi, Mozambique	Storms
10/2–7	Western Sahara, N. Mauritania	Fierce storms, destructive winds
10/10–18	S. Zaire, Zambia, Zimbabwe	Destructive storms, gales
10/20–30	Western Sahara, N.Mauritania	Storms, deluges, floods
10/22–26	S. Malawi, Mozambique	Destructive storms, floods
10/22–28	Niger, Congo, Zaire	Fierce storms, powerful winds, floods
10/23–28	S. Nigeria	Fierce storms, floods
10/24–28	Morocco, Ivory Coast, Burkina Faso, upper Egypt	Fierce storms, floods
10/29–31	S. Saudi Arabia	Storms
10/25–29	S. Tanzania	Storms, deluges
11/7–11	Ghana, Togo	Deluges
11/8–10	S. Saudi Arabia, Guinea	Deluges, floods

Dates	Locations	Events
11/22–29	Sierra Leone, Liberia, Guinea, Ghana	Fierce storms, winds, floods
11/30–12/2	S. Sudan, N. Uganda	Storms
12/2–6	Yemen, S. Saudi Arabia	Storms
12/4–10	Chad, Libya, Cameroon	Storms
12/14–16	NE Zambia, Malawi	Storms
12/21–25	Yemen, S. Saudi Arabia	Powerful storms, floods
12/22–25	SE Libya, Zimbabwe	Powerful storms, deluges
12/23–30	S. Mozambique	Fierce storms, deluges
12/27–31	SE Saudi Arabia, Chad, Cameroon, Libya	Fierce storms, deluges
12/28–31	Egypt, Eritrea, Persian Gulf, U.A.R.	Fierce storms, deluges

Bibliography

Selected Bibliography and General References

Much of the information contained in this book has been gathered from Australian and American daily media reports on the environment and natural disasters, as well as science magazines, and atlases.

Books

Goodavage, Joseph. *Our Threatened Planet*. New York: Simon and Schuster. 1978

Cornell, James. *The Great International Disaster Book: Third Edition*. New York. Charles Scribner's Sons. 1982

Ludlum, David.M. *The American Weather Book*. Boston, Massachusetts: Houghton Mifflin. 1982

Nelson, J. H. *Cosmic Patterns*. Washington D.C: American Federation of Astrologers. 1974

Ward, Bernie. *Nostradamus*. New York: Globe Communications Corp. 1997

Pearce, Alfred John. *The Text Book Of Astrology*. Washington D.C: American Federation of Astrologers. 1970

Gribbin, John and Plagemann Stephen. *The Jupiter Effect*. The Macmillan Press Ltd. 1974

Gribbin, John. *The Hole in the Sky*. Great Britain: Corgi Press. 1988

Velikovsky, Immanuel. *Earth In Upheaval*. New York: Dell Publishing. 1955

Braden.Gregg. *Awakening to Zero Point*. Washington: Radio Bookstore Press. 1997

Magazines, Periodicals, Newspapers, Atlases

Bild (Germany) "Six Planets In A Row. Should We Be Afraid?", page 10, 5/1998

The Great Geographical Atlas: Rand McNally and Company, 1991

Discover Magazine: Special Issue:–The Year In Science, 1/1999, Disney 1998

Whittingham. H. E. "The Bathurst Bay Hurricane Report." Handed to Brisbane Bureau of Meteorology, 7/1958

Television

Discovery Channel, "Santorini Volcanic eruption in Greece linked to the Exodus"

Index

LOOK FOR THE CRESCENT MOON

Llewellyn publishes hundreds of books on your favorite subjects! To get these exciting books, including the ones on the following pages, check your local bookstore or order them directly from Llewellyn.

ORDER BY PHONE

- Call toll-free within the U.S. and Canada, 1-800-THE MOON
- In Minnesota, call (612) 291-1970
- We accept VISA, MasterCard, and American Express

ORDER BY MAIL

- Send the full price of your order (MN residents add 7% sales tax) in U.S. funds, plus postage & handling to:

  Llewellyn Worldwide
  P.O. Box 64383, Dept. K414-6
  St. Paul, MN 55164–0383, U.S.A.

POSTAGE & HANDLING

(For the U.S., Canada, and Mexico)

- $4 for orders $15 and under
- $5 for orders over $15
- No charge for orders over $100

We ship UPS in the continental United States. We ship standard mail to P.O. boxes. Orders shipped to Alaska, Hawaii, The Virgin Islands, and Puerto Rico are sent first-class mail. Orders shipped to Canada and Mexico are sent surface mail.

International orders: Airmail—add freight equal to price of each book to the total price of order, plus $5.00 for each non-book item (audio tapes, etc.).

Surface mail—Add $1.00 per item.

Allow 4–6 weeks for delivery on all orders.
Postage and handling rates subject to change.

DISCOUNTS

We offer a 20% discount to group leaders or agents. You must order a minimum of 5 copies of the same book to get our special quantity price.

FREE CATALOG

Get a free copy of our color catalog, *New Worlds of Mind and Spirit.* Subscribe for just $10.00 in the United States and Canada ($30.00 overseas, airmail). Many bookstores carry *New Worlds*—ask for it!

Visit our website at www.llewellyn.com for more information.